Agricultural Growth and Economic Stability

Agricultural Growth and Economic Stability

Case Study of North Bihar

K.P. RANJAN

Rawat Publications
Jaipur and New Delhi

ISBN 81-7033-632-5

The publication of this book was financially supported by ICSSR and the responsibility for the facts stated opinions expressed or conclusions reached, is entirely that of the author and that ICSSR accepts no responsibility for them.

ICSSR Consultant
Prof. G.K. Chadha

Published by
Prem Rawat for ***Rawat Publications***
Satyam Apartments, Sector 3, Jain Temple Road,
Jawahar Nagar, Jaipur 302 004 India
Phone: 0141 651748 / 657006 Fax: 0141 651748
e-mail: rawatbooks@pinkline.net

Delhi Office
G-4, 4832/24, Ansari Road, Daryaganj, New Delhi 110 002
Phone: 3263290

Typeset by Rawat Computers, Jaipur
Printed at Nice Printing Press, New Delhi

Dedicated to
Members of my family
for their faith, courage, cooperation, sacrifice,
perpetual affection and guidance which inspired me to
higher ambition in life.

Contents

Foreword

Bihar agriculture, with its fertile alluvial soil and favourable agro-climatic condition, has been moving on the path of progress since independence, though with a slow pace. In fact, the agricultural development is predominantly dependent on infrastructural facilities created in the state under successive Five Year Plans. Despite a continuous decline in net sown area, there has been a steady growth in production of principal crops in Bihar but the increase in their productivity has not been impressive during the last 30 years. The foodgrain production is still not enough to meet the minimum requirements of the population in the state.

Dr. Kumar Priya Ranjan has made an analytical study of the agricultural growth and stability in Bihar which is being published with the financial assistance from Indian Council of Social Science Research, New Delhi at appropriate time. His attempt to look into the detailed regional analysis of cropping pattern and intensity, output growth and instability, causes of instability in production, and components of agricultural growth in Bihar is indeed praiseworthy. He used the most upto-date analytical tools for precise and reliable conclusions. I am glad that Dr. Kumar has made the study of growth in agriculture output in Bihar for the last 30 years. He has observed a significant change in the cropping pattern during the post-green revolution period, however foodgrain crops still predominate, probably due to foodgrain scarcity in Bihar. He examined the components of growth in foodgrain production and found irrigation as an important factor which contributed nearly 60 per cent to the growth of foodgrain production in Bihar. It has an important policy implication because an increase in investment in irrigation would increase agriculture production in state of Bihar. Moreover, an investment in agricultural research is also required for evolving varieties and package of practice for increasing agriculture production since the growth in foodgrain production would only be feasible through increasing productivity of principal crops.

There is a dearth of literature on regional analysis of agriculture development because there is a tendency on the part of many economists to analyse national level data of agriculture development. All-India studies are, no doubt, necessary but the regional analysis is also important if we want to formulate realistic regional strategies for agriculture development. The book brings together much needed empirical evidence on agriculture growth and instability of agriculture production in different zones of North Bihar and provides a systematic analysis of components of growth and reasons for instability in Bihar agriculture which may be used in formulating policies and strategies for agriculture development in Bihar and other states with similar agro-hydrological ecosystem and socio-economic environment. I hope, the book will stimulate further research and analysis in the field and help policy-makers in identifying critical issues of agricultural growth for faster agriculture development in backward states like Bihar.

Ajit Kumar Singh

Chairman, NAFED, New Delhi

Preface

The stability of agricultural output in the sense of elimination of large year-to-year fluctuations in output due to weather and rainfall changes is important for economy and the growth in output over time. The fluctuations in agriculture disturb the rhythm and the mutual balance among the many interdependent processes in the economy and enhance the degree of uncertainty about future that is faced by policy-makers as well as producers and consumers in the economy. Paradoxically, despite the strengths of our agriculture system, the annual variations in output around the rising trend appear to be greater in the post-independence period when compared to a few decades preceding independence. Indeed, there is an indication of some increase in the variability, particularly in foodgrains after the onset of the green revolution whereas the technology and input base of our agriculture has improved considerably. The complementarity between such input base and rainfall seems to have become stronger. Agricultural production has now become more sensitive to variation in rainfall.

The nature of factors affecting these fluctuations in such a way that they vary from year to year in a seemingly random and unpredictable manner. Measurement of contribution of different factors to growth and instability would also be important for manipulating controllable factors affecting variations in growth and instability in agriculture production. At this juncture, regional analysis of growth and instability in agriculture production along with their contributory factors need to be investigated.

In this context, the interest revolves round the nature of growth-instability relationship in agricultural production. This relationship emerges in the context of new technology, especially where agricultural productions touch optimum levels and also in the context of regional diversities, particularly in areas affected by natural calamities like, flood and drought. Hence, the present study has been aimed to understand the nature and behaviour of agricultural production along with the relationship between its growth and instability during the post-green revolution period to formulate appropriate policies/strategies for the development of agriculture in the 21st century.

The study is primarily focused on agricultural scenario of North Bihar, however the zone-wise study has also been made to have a comparative picture of regional agricultural growth and instability. The study covers the post-green revolution period that is, 1970-71 to 1993-94 which has further been categorised in three sub-periods like, Period I (1970-71 to 1979-80), Period II (1980- 81 to 1989-90) and Period III (1990-91 to 1993-44). The analysis is based on secondary data on area, production, productivity of principal crops along with different agro-economic variables; e.g., rainfall, area under HYV, fertilizer consumption, irrigated area, etc., affecting agricultural production in the project area.

A large number of studies have been conducted on growth and instability in agricultural production in general and foodgrain production in particular. Most of studies in this field dealt with the post-green revolution period and almost all of them are based on national and state level data. If rainfall and agro-climatic conditions are major factors influencing stability which vary much even within the state, regional studies with different sub-periods of the post-green revolution would generate much meaningful information on these important aspects of agricultural development. To bridge this information gap, the present study has been undertaken to analyse the nature and extent of growth in foodgrain production and to examine whether the introduction of seed-fertilizer technology has made the crop production more stable/unstable in backward region like North Bihar. The present analysis has an edge over the existing studies on agricultural growth and instability on at least four counts: (i) It examines the growth rates and instability of principal foodgrains crops separately; (ii) analysis was done for two agro-climatic zones of North Bihar; (iii) it covers the period upto 1993-94, for which revised estimates are available; and (iv) it analyses production growth and instability during three sub-periods of the post-green revolution.

The conclusions drawn by the present study seem to be more reliable because various sophisticated statistical tools have been employed to reach at relevant conclusion. There is hardly any publication which covers the agricultural growth dynamics starting from cropping pattern to growth instability and factors contributing to growth. It is worth pointing out that pioneer effort has been made to investigate issues mentioned above in the backward economy. Publication of this book would have immense use for academicians, policy-makers and administrators. It may also be useful for national and international organisations dealing with regional agricultural growth and development.

Words are inadequate to express my deep sense of gratitude

and devotion to the reverend chairman of the advisory committee Dr. R.K.P. Singh for keen interest, concrete suggestions, enlightened guidance and painstaking help during research work, I owe him much. I am extremely obliged to Dr. B.N. Verma, University Professor & Head, Department of Agricultural Economics, RAU, Pusa for his valuable suggestions and for providing necessary facilities during the course of present investigation.

I am very much thankful to faculty members of the Department of Agricultural Economics, RAU—Dr. J.N. Choudhary, Dr. R.N. Yadav, Dr. A.K. Choudhary, Dr. B.B. Singh, Dr. J. Thakur and Dr. C.P. Yadav—for their help and suggestions. I would also like to thank Dr. S.P. Singh and Dr. A. Bhattacharya for their unstinted support, guidance and constant encouragement in finalising the draft of the manuscript.

I also put on record my appreciation to faculty members and other staff of the DNS RICM, Patna for their cooperation and help.

Finally, I shall be failing in my duty if I do not thank the ICSSR for the financial assistance and also Rawat Publications for having gracefully undertaken the publication of the book.

Kumar Priya Ranjan

1

Introduction

Agriculture still holds key of Indian economy since it contributes about one-third to the gross domestic product and provides employment to nearly two-third of working population. The entire national economy behaves to the tune of agriculture sector. The general price level tends to decline in the year of bumper agricultural harvest whereas abnormal agricultural years experience a sharp rise in prices of even industrial commodities.

The agriculture sector has performed impressively in terms of increasing productivity and intensity of cultivation. It is by no means a small achievement that the foodgrain availability per capita at country level has increased to high level over the last four and a half decades, in the face of an assault of population pressure and declining per capita land availability (from one-third of a hectare in the 1950s to less than one-seventh in the early 1990s). In the same period, irrigation availability has doubled and cropping intensity has increased significantly at the country level. This could be possible due to cutting edge of science, coupled with fast adoption of technology by the farmers, and above all, the government decision to accord a high priority to agriculture by making large plan investments in infrastructure. As a result of technological developments and their effective dissemination, wheat production rose from 6.46 million tonnes in 1950-51 to 65.80 million tonnes in 1994-95. The productivity increased from 663 kg/ha to 2,559 kg/ha during the period. Rice production also increased from 20.58 million tonnes in 1950-51 to 81.80 million tonnes in 1994-95. Production of oilseeds increased from 10 million tonnes in 1982-83 to 21.48 million tonnes in 1996-97. Unlike in the past, crisis situations of drought or flood are now considerably well managed without any panic or large-scale imports. On the contrary, India faced successfully the worst drought of the century in the year 1987-88 which speaks of resilience of Indian agriculture.

During the post-green revolution period, area under major cereal crops would ordinarily be changing substantially from year to year. The major technological improvement in agriculture sector is accompanied by an upward trend, both in its area and yield. Factors promoting growth of agriculture are usually the factors, bringing about change in crop composition of agriculture.

An important dimension of agricultural growth is the spatial variations which is very relevant in a vast country like India, having a wide range of crop-soil-weather conditions. During the last decade, the states recording above national average annual growth in foodgrain output were predominantly Punjab, Madhya Pradesh, Bihar, Haryana, Uttar Pradesh and West Bengal. The states of Gujarat, Kerala and Maharashtra had witnessed negative rates of growth as compared to the preceding decade (1971-80). While the rate of total foodgrain output for the country as a whole has increased from 2.31 per cent (1967-68 to 1977-78) to 2.68 per cent (1977-78 to 1988-89). The considerable year to year variations in foodgrain output occur on account of inherently adverse agro-climatic situation persisting in certain areas of the country (Hanumantha Rao and Gulati, 1994). The post-green revolution period till the end of 1980s witnessed reliance on high productivity regions and productivity augmenting techniques, which were often short-term resource exploitation measures leading to instances of water-logging problem, soils and other resource related problems. Growth in agriculture would, thus, have to be achieved through the adoption of a disaggregated approach which gives due attention to the inherent differences of environment. The most prominent contributors to the all-India agriculture output have been the states of Uttar Pradesh, Maharashtra, Andhra Pradesh, Tamil Nadu, Punjab and Rajasthan. These states together contributed 54.72 per cent to the agriculture gross domestic product (GDP) of the country in 1989, a change of only 3 per cent from 51.79 per cent contribution in 1980. The states of Haryana, Punjab, Maharashtra, Rajasthan and Karnataka have considerably high index of per capita output as compared to the country, indicating that agriculture plays a prominent role in generating income for the rural population.

The stability of agricultural output, in the sense of elimination of wide year-to-year fluctuation in output due to weather and rainfall changes, is an important aspect for sustainable growth in output over time. The fluctuations in agriculture disturb the rhythm and the mutual balance among the interdependent processes in the economy and enhance the degree of uncertainty about future that is faced by policy-makers as well as producers and the consumers in the economy (Rao, 1986). Paradoxically, despite the strength of our agriculture system, the annual variations in output around the rising trend appear to be greater in the post-independence period when compared to a few decades preceding independence. Indeed, there is an indication of some increase in the variability, particularly in foodgrains after the commencement of the green revolution whereas the technology and input base of our agriculture has improved considerably. The complementarity between such input base and rainfall seems to have become stronger. Agricultural production has now become more sensitive to variation in rainfall.

The nature of factors affecting these fluctuations is such that they vary from year to year in a seemingly random and unpredictable manner. Measurement of contribution of different factors to growth and instability would also be important for manipulating controllable factors affecting variations in growth and instability in agriculture production. At this juncture, regional analysis of growth and instability in agricultural production along with their contributory factors need to be investigated.

North Bihar with 1.62 per cent of the landmass, supports nearly 4.61 per cent of population of the country. The region has almost the lowest foodgrain productivity (1,266 kg/ha) and per capita land (0.12 ha/man) supports comparatively larger number of per hectare population (8.27). It has a typical agro-climatic situation and landscape. Rainfall varies from 1,050 mm to 1,700 mm per annum, however, its distribution is uneven. The cropping pattern is cereal dominated and the foodgrain production increased from 29.13 lakh tonnes in 1970-71 to 57.51 lakh tonnes in 1993-94. Despite the substantial increase in foodgrain production, the area has been deficient in foodgrains, particularly, in the years of drought. There were four drought years and three bumper years in North Bihar, during last 24 years which registered decline in foodgrain production up to 17 per cent, during drought period and spectacular increase up to 39 per cent in bumper years (Appendix I). Hence, it clearly indicates the role of weather, in causing instability due to poor infrastructure in the region. Moreover, other factors must have been affecting the growth and instability in agriculture production which needs immediate investigation.

In order to examine agricultural growth and instability and their different components, during the post-green revolution period, the present study was planned to analyse the nature and extent of agricultural development in backward region like, North Bihar. It also aims at studying whether the new technology has made the crop production more stable or unstable? The study would have far-reaching implication for the planners, policy-makers and promoters of change for formulating strategies for agricultural development in backward economy. The detailed objectives of study are as follows:

1. To examine the changes in cropping pattern during the post-green revolution period.
2. To assess the area, production and productivity changes in North Bihar, during the post-green revolution period.
3. To study the growth rates in area, production and productivity of principal crops, during period under study.
4. To measure the extent of instability in production of principal crops as well as its relation with growth.
5. To measure the contribution of different components to agricultural production/growth.

Scope of the Study

The study deals with agricultural growth and instability in agriculture production of North Bihar which is quite different from other parts of the country with respect to agro-economic situations. Hence, the conclusion of the study would only be helpful in formulating policies and strategies for agricultural development in the regions having typical agro-climatic situations of North Bihar. The findings of the study could be replicated in similar agro-climatic situations, in general and the other parts of

Bihar, in particular. Besides, it will provide useful guidelines and directions to similar type of studies in other parts of the state and the country. It may provide useful information for formulating hypotheses for further research in agricultural growth, either with the help of cross-sectional data or time series data. The findings of the study relating to components of production/growth of foodgrains could be utilised for assigning priority in agricultural development programmes in the project area.

Layout of the Book

The book is presented in eight chapters. The first chapter is an introduction with objective and scope of the study. Review of literature is presented in the second chapter. This is followed by the third chapter which deals with research methodology, consisting of period of study, analytical tools and limitations of the study. The fourth chapter covers the agro-economic background of the project area comprising demographic features, soil and climate, infrastructure facilities, and level of input used. The findings and discussions are presented in the fifth, sixth and seventh chapters. In the fifth chapter, the cropping pattern, decade-wise change in area, production and productivity of principal crops have been discussed, whereas the growth and instability in production of principal crops has been examined in the sixth chapter. The components of production/growth are described in the seventh chapter. The last chapter contains a brief summary and conclusions of the investigation, followed by bibliography and appendices.

References

Hanumantha Rao, C.H. and Gulati, A. (1994). "Indian Agriculture: Emerging Perspectives and Policy Issues", *Economic and Political Weekly*, 29 (53): A-158.

Rao, M.V. (1986). "Agricultural Production: Pace and Pattern of Growth", *Indian Agricultural Development since Independence*. Oxford & IBH, New Delhi, p. 95.

2

Studies on Growth and Instability in Agricultural Production in India

Indian agriculture has usually been characterised as a 'gamble in monsoon'. A great deal of uncertainty in crop prospects still exists, as monsoon plays a decisive role in determining agricultural output and its failure results in widespread famine and misery. In this context, interest revolves round the nature of growth-instability relationship in agricultural production. This relationship emerges in the context of new technology, especially where agricultural productions experience troughs more often than peaks, and also in the context of regional diversities, particularly in areas visited by natural calamities like, flood and drought. Hence, the present study has been planned to understand the nature and behaviour of agricultural production along with the relationship between its growth and instability. To understand the intricacy of the objectives and methodologies of identical studies conducted in Indian context, the relevant empirical research studies are reviewed under the following sub-heads:

(a) Change in Cropping Pattern.

(b) Growth and Instability.

(c) Factors Affecting Agricultural Production.

Change in Cropping Pattern

A dynamic change has been witnessed in agricultural scene in our country, particularly during post-green revolution period. The technological advancement in crop varieties and other yield increasing factors of production influenced the farmer's behaviour which got reflected in the changing cropping pattern from cultivation of low value crops to high value crops in most of the regions. In the present section, an effort has been made to review the relevant research studies conducted on cropping pattern in different parts of the country.

Chawla *et al.* (1973) observed that as a result of state tubewell irrigation, a significant shift took place in the cropping intensity and cropping pattern in the project area (Amritsar district, Punjab). The cropping intensity increased from 126 to 164. Among cereals, wheat and paddy gained an area to the extent of 50.90 and 189.10 per cent respectively, while the area under barley, maize and millets showed a decline by 14.30, 13.40 and 210.70 per cent respectively. Area under pulses also reduced by 53.33 per cent. This shows that less remunerative crops were replaced by more remunerative crops due to irrigation effect.

Ray *et al.* (1985), while analysing state-wise data to assess the impact of changes in cropping pattern on farm income, observed that the spread of new technology in agriculture has accelerated disparities in income distribution among states. It is, therefore, desirable to initiate steps to correct such imbalances in growth with equity. The study reveals that agricultural income gap among the states has widened mainly because of the nature of cropping pattern in different states. The states which have increased their area under wheat, rice, sugarcane, etc., have been able to increase their income much more than other states where low income crops predominate. This highlights the imperative for developing yield increasing technology appropriate to such areas and crops like oilseeds, pulses, *bajra* and *jowar*. In addition, there is a need to develop suitable marketing infrastructure, institutions and appropriate pricing policies so as to ensure high income to the producers of these crops in the absence of which further increase in production may not be forthcoming.

Venkataraman and Prahaladachar (1981) estimated and analysed the growth rates in area, yield and output of major crops in six states including Bihar for the period 1950-51 to 1974-75, and also examined the impact of growth rates on cropping pattern in these states. In Bihar, the study of cropping pattern indicated that within the cereals like rice, maize and wheat gained in area whereas barley lost in area during 1950-75. In the cropping pattern, the relative importance of wheat and maize improved from 6 to 15 per cent and 6-18 per cent respectively. The study concluded that the cropping pattern shifted in favour of cereals at the cost of pulses and oilseeds. In Bihar, the output growth rate during the period 1950 to 1975 was highest for wheat (9.5%), followed by maize (3.7%). It was also observed that nearly 75 per cent of the growth in crop output was due to contribution made by yield growth whereas the contribution of area growth and changes in cropping pattern was small.

Bhat *et al.* (1989) studied "Crop Concentration and Cropping Pattern in Jammu & Kashmir state" and found that on relative basis the area allocation towards the oilseeds, fruits and vegetables and non-food crops has increased from 3.01, 3.12 and 2.89 per cent of the total cropped area in 1970-71 to 5.94, 5.35 and 3.46 per cent respectively, in 1983-84.

Lal and Singh (1994) observed that technological breakthrough in Haryana agriculture has brought about significant shifts in the production, productivity and cropping pattern. The results further revealed that agriculture in the state has undergone significant change. However, the benefits of such change have not been shared equally across the crops and regions in the state. This resulted in an impressive growth in the production of the superior cereals like paddy and wheat while the coarse cereals and non-foodgrain crops except oilseed have not received due attention in the growth process. This invites special attention of the researchers and policy-makers towards the crops along with equitable exploitation of the irrigation resources for sustainable growth in agriculture. Thus, changes signify that there has been a shift from cereal economy to market economy which is a very healthy sign for the development of agricultural sector in Jammu & Kashmir.

Growth and Instability

Krishna and Tyagi (1973) in their study, "The Growth Rates of Area, Production and Productivity of Wheat in Uttar Pradesh for the Period 1950-51 to 1970-71", observed that the area under wheat, its production and productivity increased by 2.77 per cent, 6.72 per cent and 2.51 per cent respectively during 1950-51 to 1970-71. An effort has also been made to find out the growth rate of area, production and productivity of wheat during the periods 1950-51 to 1964-65 (before green revolution) and 1965-66 to 1970-71 (post-green revolution). It was observed that the growth rates of area, production and productivity which were 1.36, 2.43 and 0.88 per cent respectively, during the period 1950-51 to 1964-65, increased to 2.77, 6.77 and 2.51 per cent during 1965-66 to 1970-71, showing considerable impact of green revolution on wheat cultivation in the state. The comparative analysis of growth rates of area, production and productivity of wheat in Uttar Pradesh as compared to Punjab, Haryana and all

India, indicated that Uttar Pradesh registered the lowest growth rates in area, production and productivity as compared to Punjab, Haryana and all India figures.

Rath (1980) examined the performance of agricultural production in India. An exponential trend function was used to estimate production, area and yield rates for the period from 1949-50 to 1977-78. The analysis indicated that the total agricultural production of India grew at an average rate of 2.38 per cent per year during the years 1955-56 to 1978-79. The rate of growth was found to be somewhat higher during the 10 years ending 1964-65 than during the subsequent 13 years when it was only about 2.42 per cent. It was also observed that the sustained growth of foodgrains was entirely due to cereals. Wheat recorded a growth rate of 3.9 per cent and rice of about 3 per cent. In the post-1965 period, with the advent of new High Yielding Varieties (HYVs), the growth rate of wheat almost doubled to 7.06 per cent, of this about half was due to increase in area and other half due to increase in productivity of the crop. However, wheat production which increased at the rate of about 11 per cent during 1964-72 registered only 3 per cent growth rate during seventies. But this was not surprising as high yielding wheat had already reached very large areas and achieved economically high level of yield. In Bihar, the annual compound growth rates of area, production and yield of wheat were observed to be 2.81, 5.65, 2.95 per cent and 3.42, 8.31, 4.04 per cent during the periods 1949-50 to 1977-78 and 1964-65 to 1977-78 respectively.

Mehra (1981) while studying instability in Indian agriculture observed an increase in the standard deviation of production of all crops by 75 per cent and of foodgrains by 65 per cent during the decade from 1967-68 to 1977-78 over that of the 1950s to mid-1960s but the mean production of all crops including foodgrains increased by only 47 per cent during the period. There was even an increase in the coefficients of variation of production of foodgrain crops during the period.

Growth in agricultural production is increasingly based on yield, and the new seed-fertilizer technology is making an overwhelming contribution to growth in yield. Though it is difficult to identify and even more difficult to quantify the causes, the evidence highlights the association between increase in yield variability and the use of HYVs and larger applications of fertiliser per unit of land. When seed-fertilizer technology is combined with assured irrigation, the tendency for variance to increase is neutralised because assured source of irrigation is likely to result in an increase in yield much higher than the increase in yield variability.

Wheat and rice crops where application of new technology has been widespread show only a small increase in absolute yield variability. The difference may well lie in the fact that both wheat and rice are highly irrigated crops. Among non-foodgrains, absolute and relative yield variability of sugarcane and relative yield variability for potato were somewhat lower in the second period. Again irrigation may contribute to stability of all crops. Sugarcane has the highest percentage of its area under irrigation.

Singh (1981) examined the imbalances in agricultural growth and found that the uneven growth rate of individual crop has led to the regional imbalances in the rural prosperity. Even for a crop like rice which is the most important cereal crop of the country, the growth rate has not been uniform in the different states. Since agriculture is the main source of income of rural population, imbalance of its growth in different regions has led to the imbalances in income and levels of living.

Bhagat (1982) in his study on "Growth Rates of Output of Coarse Grains and Pulses in a Backward Economy: A Study of Temporal-Spatial Variations in Chotanagpur", examined the magnitude and direction of changes in area, yield rate and output of major coarse foodgrains and pulses in relation to their competing crops in the districts of Chotanagpur region in Bihar in the two sub-periods, i.e., pre-HYV (1956-57 to 1965-66) and post-HYV (1967-68 to 1976-77) periods along with the causes underlying these changes and variations. It was found that there was no marked difference in the growth rates of crops grown during the rabi season in the pre-HYV period but the growth rates of area and output of barley and gram were found to be much lower than that of wheat in the HYV period, indicating the farmers' preference for growing wheat as against barley and gram, however, the reasons for comparatively higher growth of wheat area and output were not investigated.

Gangwar and Pandey (1982) examined the stagnation in production of pulses and pointed out that the trends of total pulses with respect to area, production and yield were statistically non-significant, reflecting the stagnation in production of pulses in the country. In spite of higher prices of pulses, wheat and other HYV cereals have become more profitable than pulses due to their substantially higher and more stable yields. This is also so due to shift of good land from pulses to HYV cereals. The study was planned to cover only pulse crops.

Hazell (1982) made an attempt to estimate the "Instability in Indian Foodgrain Production" and reported that the coefficient of variation of total cereal production was 5.85 per cent, during the period 1967-68 to1977-78 when measured around the trend. This was nearly 50 per cent larger than the coefficient of variation (4.03%) during the period 1954-55 to1964-65. It is still quite modest by international standards but these fluctuations can cause substantial price instability because of the high proportion of production retained by farm families.

The analysis confirms that changes in variability of yield of individual crops within states have been an important contributor to the increase in the coefficient of variation of total cereal production. However, increase in crop yield within states has been far less important for the instability of all India production than increase in inter-state yield co-variances and increase in area yield co-variances both within and between crops and states. Increase in yield variance only accounts for a small part of the increase in the yield and area yield co-variances. More important sources of increase have been a definite change towards more positive correlations between the yield fluctuations of different crops in the same and in different states and an increase in the variability of crop areas sown which now are also more positively correlated with yields. The sources of these fundamental shifts in patterns of association are difficult to identify, but they probably have less to do with the improved seed/fertilizer based technologies than with changes in weather patterns and the more widespread use of irrigation and fertilizers at a time when the supplies of these inputs are not reliable.

Singh *et al.* (1982) tried to study the "Growth Rate and Trend in Area, Production and Productivity of Oilseed Crop in Haryana" for the period 1966-67 to 1980-81 and found that the area under oilseeds has increased to 310 thousand ha from 212 thousand ha during the period under report. The productivity has also increased to 607 kg per ha from 440 kg per ha during the period. The compound growth rate of area was found to be 0.5 per cent, however, the compound growth rates of production and productivity were 2 per cent and 1.6 per cent respectively and found significant. The comparatively higher variability was observed in the production followed by area. The coefficient of variation for productivity was only 19.25 per cent. In the increase of oilseed production, the contribution of area was 45.93 per cent and that of yield was 54.07 per cent.

Chandra *et al.* (1983) conducted a study on "Vertical Growth for Agriculture in Uttar Pradesh" and observed that as against a growth rate of only 1.25 per cent per annum in the area of wheat, during Period I (1950-51 to 1964-65) in the state, the growth rate increased to a higher level of 4.54 per cent per annum, during Period II (1965-66 to 1977-78). The overall growth during the period 1950-51 to 1977-78 worked out to 2.66 per cent per annum. As regards the compound growth rate of acreage under rice, the growth rate was lower in Period II as compared to Period I.

Wheat production in the state increased at a much faster rate during Period II as compared to Period I and as such the state registered a growth rate of 6.45 per cent per annum dúring Period II. The rapid growth in the production of wheat in the state clearly indicates a significant departure in the trend line of production from 1965-66 and onwards. Rice production in the state rose at the rate of 6.14 per cent per annum, during Period II which was higher than that of Period I.

Productivity growth of wheat in the state during first period was only 0.72 per cent per annum which increased to 2.55 per cent per annum during Period II. The overall growth during the period 1950-51 to 1977-78 was 2.3 per cent per annum. In spite of decrease in the growth of acreage under rice during period second, its productivity in the state increased at the rate of 5.15 per cent per annum.

Grewal and Rangi (1983) made an attempt to estimate the growth of Punjab agriculture in relation to productivity increase and response to the use of fertilizer and observed that productivity increase was largely the outcome of increased irrigation and fertilizer use. It was further estimated that one kilogram of fertilizer nutrients yielded about 8 kg of wheat grains during the period 1967-68 to 1976-77 when the average use of fertilizer for wheat crop was about one-third of the recommended dose. During the succeeding period ending 1981-82, this response was only 5.6 kg of wheat per kg of fertilizer nutrients applied. This was the case when fertilizer use was two-third of the recommended level.

Sirohi *et al.* (1983) while conducting study on "Trend of Agricultural Productivity and Production in India" did not support the view that the trend in foodgrain production or total agricultural production has been stagnant or has declined, during the recent period. They examined the trend of production and productivity of wheat per hectare which has significantly increased during the study period, i.e., 1970-77 to 1981-82. The productivity of some crops like *bajra*, pulses and gram has been stagnant or decreased slightly. Crops like rice and cotton have shown a rising trend. The productivity of total foodgrains per hectare has also increased during recent years mainly due to rise in the productivity of wheat crop. The study also indicated that the yearly fluctuations in agricultural productivity and production were so large that they completely overshadowed the general trend of rise and fall of productivity and production.

Bhat *et al.* (1986) made an attempt to study the growth rate in production of major crops in Jammu & Kashmir state for the period 1970-71 to 1983-84 and found that the compound growth rate of oilseeds had shown the highest growth rate of 4.85 per cent followed by rice and wheat with 1.85 and 1.19 per cent respectively which had been observed to be significant at one per cent level of probability. However, the compound growth rate of area under pulses indicated non-significant trend, i.e., 0.30 per cent but the compound growth rates of production were found to be highly significant in case of rice (3.80%), wheat (2.78%) and oilseeds (10.01%). But in case of pulses, production growth rate has been observed as non-significant, i.e., 0.62 per cent.

George and Mukherjee (1986) in their work on "A Disaggregated Analysis of the Growth Performance of Rice in Kerala", found that the growth rate of area, yield and production fluctuated considerably across the districts, over seasons and over time. In spite of the overall decline in area for certain seasons, the growth rate of yield for the Period II (1975-76 to 1983-84) was much higher than that of the Period I (1960-61 to 1974-75), particularly during autumn and winter seasons, but during summer the growth rate of yield remained more or less constant for both the periods. The growth rate of rice production has declined during the Period II but the decline was comparatively higher in case of summer rice.

The proportion of irrigated area (both HYV and non-HYV) turned out to be a non-significant variable in explaining paddy yield. While the proportion of area under HYV had a significant influence on yield during autumn and winter but it was found non-significant during summer. Because of this difference in the annual data, HYV area turned out to be unimportant in explaining rice yields. Inter-district variations in paddy yield had been significant in many situations but inter-seasonal variations were not significant.

Singh and Gangwar (1986) made an attempt to examine "Trends and Variability in Area, Production and Productivity of Coarse Grains and Pulses in Haryana" for the years 1966-67 to 1980-81" and found that the growth rate of area under green gram and lentil declined at the rate of 10.67 and 2.66 per cent per annum respectively which were found to be statistically significant. The area under gram and black gram also declined at the annual rate of 1.65 and 0.18 per cent, respectively.

There had been an increase in the yield of *jowar* and barley at the rate of 1.10 and 1.60 per cent per annum respectively. The yield of maize declined at the rate of 1.19 per cent per year followed by *bajra* (1.05 per cent) which was statistically non-significant. The yield of green gram, black gram and lentil increased while the yield of gram declined but it was non-significant. The yield of green gram, black gram and lentil had growth rates of 6.9, 4.7 and 2.3 per cent per annum respectively. The productivity trend of pulses was near to zero indicating stagnation in the yield of total pulses.

Mahendra Dev (1987), while examining the "Growth and Instability in Foodgrain Production", found that the inclusion of rainfall index in the trend equation has improved the value of coefficient of multiple determination (R^2) in most of the states which indicates that rainfall was able to explain the major part of the variation in foodgrain production. Weather adjusted growth rates were higher in some states and lower in some others as compared to unadjusted growth rates. The rate of growth in foodgrain production was ahead of population growth only in five states, namely, Maharashtra, Punjab, Uttar Pradesh, Haryana and Andhra Pradesh while it was trailing behind in rest of states.

Trends in instability for four crops aggregates (rice, wheat, coarse cereals and pulses) show that instability for rice, coarse cereals and pulses declined in some states and increased in some other states. On the contrary, it declined in most of the states for wheat. It was shown that differences in quahtity and quality of irrigation might be the major factor that influenced inter-state variations in growth and instability in foodgrain production.

Prasad (1987) in "A Study on Trend in Area, Production and Productivity of Pulses in Bihar" during 1960-61 to 1989-90 reported that the area, production and productivity of gram and lentil were consistently higher in Zone III, comprising old districts of Patna, Gaya, Shahabad, Bhagalpur and Monger, while that for *arhar* they were higher in Zone I consisting of old districts of Saran, Champaran, Muzaffarpur and Darbhanga.

Growth rate analysis indicated that the compound growth rates of area and production of pulses had, by and large, declined all over the state, except that for production of gram in Zone VI (0.11%), production of *arhar* in Zone V (0.37%) and production of lentil in Zone I (1.23%), Zone II (4.40%) and

Zone II (6.49%). A positive growth rate in area was found in lentil only in Zone II (3.34%) and Zone IV (6.25%). Productivity of all the pulse crops under study had positive growth in all the zones of Bihar during the period under study. The factors like rainfall and area under irrigation had significantly affected the production of pulse crops only in some of the districts of the states.

Hanumantha Rao *et al.* (1988) conducted a study on unstable agriculture and droughts and revealed that the growth of foodgrain output as a whole has been around 2.7 per cent per annum since the beginning of planning in India and there is no evidence of an acceleration or deceleration in the growth rate between the pre and post-green revolution periods. Wheat and rice provide a contrasting picture of growth, the former showing an acceleration and the later a deceleration in the post-green revolution period. In the more recent years, however, growth rate of rice seems to be picking up, whereas that of wheat seems to be sliding down.

There are wide disparities in growth rate of foodgrains output in different states. These disparities seem to have widened in the post-green revolution period when much smaller number of states experienced above average growth rate than during the 1960s. The variation in the performance of individual states over the pre and post-green revolution periods seemed to be as significant as inter-state variation in performance within each of the two periods.

The fluctuations in the output of all crops, except wheat, have increased significantly in the post-green revolution period (1966-85) when compared to the pre-green revolution period (1950-65). The instability in the case of foodgrains as a whole or all crops put together, is significantly lower than for individual crops, indicating that the observed increase in instability in the case of foodgrains cannot be attributed to green revolution as such. The instability in agricultural production has increased in the post-green revolution period on account of rise in the sensitivity of output due to variations in rainfall. This rising vulnerability of agricultural output,especially of foodgrain output to drought,is traceable to the high complementarity of new seed-fertilizer technology with water and inadequate expansion of irrigation facilities. Although area under irrigation has increased from about 17 per cent of cultivated area in early 1950s to around 30 per cent in mid eighties, a good part of this irrigation is itself dependent on rainfall. In this sense, the uncertainty of irrigation has been increasing in the recent period. The instability in crop production would have been substantially lower if there was faster increase in infrastructure, like public irrigation and rural electrification.

Pal and Sirohi (1988) in their investigation on "Sources of Growth and Instability in the Production of Commercial Crops in India", considered two time periods, viz., 1950-51 to 1964-65 and 1967-68 to 1983-84 for the analysis. They observed a substantial decline in the rate of growth of area caused the deceleration in the production of all commercial crops (such as a sugarcane, groundnut, cotton and jute) except potato in the post-technology period. In potato, an appreciable increase in yield led to a comparatively higher growth in production. The rate of growth in yield increased in groundnut, cotton and jute but decreased in sugarcane.

The deceleration in production was accompanied by the increase in production instability in groundnut, jute and potato in the period 1967-68 to 1983-84. However, the instability in production declined marginally in sugarcane and cotton. An increase in the instability in per hectare yield was observed for groundnut and jute and a decrease in yield instability for sugarcane and potato. In case of sugarcane, a highly irrigated crop, yield instability was much lower than production instability in all the states. In the majority of the states, the coefficient of variation of production and yield were in the range of 10 to 20 per cent. The production and yield of groundnut in Gujarat and cotton in Maharashtra were highly unstable in both the periods. Further, the production of groundnut in Maharashtra and cotton in Andhra Pradesh became highly unstable mainly due to a substantial increase in yield instability for groundnut in Tamilnadu and potato in Punjab, the instability in production increased markedly without much increase in the instability in yield. Contrary to this, the production of sugarcane in Tamil Nadu and jute in Bihar showed impressive reduction in the instability over time.

Malik *et al.* (1988) conducted a study to find out linear and compound growth rates of area, production and productivity of wheat in Haryana for the period from 1962-63 to 1983-84. It was observed that during the period 1962-63 to 1985-86, the area under wheat in Haryana increased by 170.54 per cent, production increased by 545.82 per cent and yield per hectare increased by 138.73 per cent. Such a tremendous increase in production of wheat in the state was due to combined effect of increased acreage and productivity. The compound growth rates of area, production and productivity of wheat for 1960-61 to 1983-84, increased significantly at an average annual growth rate of 4.83 per cent, 9.65 per cent and 7 per cent respectively.

Bandyopadhyay (1989), while conducting a study on "Growth and Instability in the Production of Main Cereal Crops in West Bengal and Punjab-Haryana", reported that the elaborate network of irrigation in Punjab-Haryana, independence of the vagaries of the monsoon/rainfall along with the development of private tubewell irrigation on a substantial scale, particularly, during post-green revolution period, may have effectively narrowed down the year-to-year fluctuations in the production of wheat. Rice crop has also been introduced in the states which grew at a much faster rate compared to that in West Bengal. Irreparable damage done to the ancient irrigation network in Bengal during the colonial rule on the other hand, has not been compensated by the development of irrigation during the era of planned development since independence in West Bengal. Irrigation has been the main technological constraint for agricultural production in West Bengal.

Singh (1989) in his study on "Agricultural Instability and Farm Poverty in India", used the coefficient of variation statistics along with the time series data of major crops for the period 1971-1986 and observed that the yield variability in case of rice crop was the highest, i.e., 29.14 per cent for Punjab followed by the states of Rajasthan, Uttar Pradesh, Andhra Pradesh, Orissa, Gujarat, Bihar, West Bengal, Himachal Pradesh and Jammu & Kashmir. It was the lowest, being 6.11 per cent, for the state of Assam. It may be noted that the states having the highest yield variability were found to be having the highest average yield per hectare. The higher average yield and higher yield variability in the case of non-traditional rice growing states are due to intensive cultivation and increasing area and yield of rice. In case of wheat, the states having higher yield variability had lower average yields in general. Hence, it may be said that the wheat is a more assured and stable crop than rice.

Maize crop showed the highest yield variability, i.e., 41.66 per cent in Tamil Nadu, followed by Gujarat (33.11%), Maharashtra (30.33%), Rajasthan (29.14%), Uttar Pradesh (28.16%) and Andhra Pradesh (23.61%). It was the lowest in Assam, being 3.82 per cent. However, this crop showed some variability in the states having higher average yield per hectare.

There was not much inter-state variation in yield variability of gram. The major cotton growing states such as Punjab, Haryana, Gujarat, Tamil Nadu and Rajasthan showed greater yield stability. In case of sugarcane, the states like Tamil Nadu, Maharashtra, Madhya Pradesh, Karnataka and Andhra Pradesh, showed the stability in sugarcane production because of having lower value of coefficient of variation and comparatively higher per hectare yield.

Chattopadhyaya (1991) in his study, broadly suggests that during the last three decades West Bengal's rice economy was subjected to technological changes with positive effects on area, yield and production of rice. It cannot be said to have been stagnant over these years. But, it has also been uneven spatially (in the districts) and temporally. The diffusion of the new technology is widely variant among districts and seasons. Increase in variability accompanied the growth of yield and output, thereby increasing the risk of rice cultivation which limits the acceptability of the technology among risk minimising farmers. The yield variability seems to be related positively to rainfall variation and negatively to increase in irrigation and fertilizer application, however, the relation is not very clear.

Deshpande (1991), while determining the growth-instability nexus in Maharashtra, found a strong negative relationship between growth and instability in pre-new technology period at state level across crops which changed to a positive but non-significant association after the advent of new technology. Region-wise analysis over the three decades revealed that the relationship emerged stronger in the case of non-drought prone (NDP) districts than that of drought prone (DP) districts. The results indicated instability as a companion of growth for DP districts and not so in the case of NDP districts. The low instability-high growth combination was observed in case of DP districts whereas a high instability-low growth combination was found for NDP districts.

Pal and Sirohi (1991) made an attempt to study the adoption of HYV seeds and growth and instability in crop production in India and reported the changes in the base (mean area and mean yield), yield variability and simultaneous instability (variance). The relative instability (CV) in wheat cultivation declined in more irrigated areas. The larger contribution of interaction terms indicates that the simultaneous change in area and yield further accentuated the production instability. Individually, yield variability was an important source of instability in most of the crops. The changes in yield might have caused the changes in area and this led to higher area-yield co-variability. The cultivation of HYVs in less endowed area and extension of cultivation to marginal and sub-marginal land might have caused higher production instability. This problem was more apparent in coarse cereals, pulses and oilseeds. Therefore, it could be concluded that the interplay of agro-climatic factors and new technology led to increase in production instability in India and HYVs can be held partially responsible for this.

Venkata Rao (1991) conducted a study on the production of 22 crops in the north-eastern region and revealed that for major crops like rice, other pulses, rapeseed and mustard, sugarcane, etc., stagnancy has set in. There is accelerating trend in the production of potato and relatively minor crops like *tur* (red gram) and groundnut. On the whole, principal crop groups exhibited less fluctuations in yield per hectare. The major contribution to the growth of crop output in agriculture came from area expansion, the contribution of yield increase was only one-third and the changes in cropping tended to lower the growth of crop output in the entire north-eastern region.

Ray (1991) studied the instability in Indian agriculture and found that the turning point, both in rice and wheat production growth rate, appeared to have occurred in the early 1970s. Growth rate in rice production which was falling, started rising at an accelerated rate from around 1971. For wheat, the growth rate which was increasing started falling at a declining rate from around 1973. The increased use of HYVs of seeds and modern purchased inputs for accelerating crop output growth rate have made crop production more sensitive to weather and price changes. On the basis of empirical results, it has been concluded that the production instability is an inevitable consequence of rapid agricultural growth and there is little that can effectively be done about it. The contrasting pictures obtained from rice and wheat production data illustrate the point. With rapid growth, stability can be achieved if the environment for crop production is brought under human control. Even with a slower growth rate, production can be made more unstable by changing prices and other controllable factors. The extent of instability is caused by uncontrollable natural factors, its intensity can be progressively reduced by controlling the factors that cause instability.

Singh and Gangwar (1991), while conducting study on "Instability in Cereal Production in Haryana", observed that aggregate production instability is an inevitable outcome of rapid agricultural growth and the little can effectively be done about it. However, policies to stabilise the supplies of fertilizers and effective irrigation facilities might be effective to reduce the instability of cereal production in Haryana.

Vaidyanathan and Mukherjee (1991) in their study on "State-wise Functional Analysis of Agricultural Growth" during the Period 1952-53 to 1976-77, reported that the proportion of variation in total foodgrain output explained by the model ranges from 62 to 94 per cent, it exceeds 80 per cent only in 5 cases out of 12 considered. While judging the strength of the model, it needs to be borne in mind that the data have both gaps (e.g., irrigation in Orissa and West Bengal) and errors (sampling and non-sampling ones in yield estimates) and hence assumptions had to be made in constructing series on fertilizer consumption. Under these circumstances, very high R^2 would indeed be rather doubtful. Rainfall and trend explain more than

80 per cent of the area variation in 9 out of the 12 states. Rainfall contribution to the area variation is high, precisely in those states where the overall explanatory power of the model is land.

Haffis *et al.* (1994) conducted a study on growth-instability of sorghum and pearl millet production and reported that the sorghum area and production increased at a much faster rate in the pre-green revolution period, however, the yield recorded a faster growth rate in the post-green revolution period. In case of pearl millet, the significant positive growth rates of production and productivity in the earlier period became insignificant in the later period. The area of sorghum and pearl millet showed negative growth rate during post-green revolution period.

The instability in production and productivity increased in the Period II. The increase in instability was marginal in case of sorghum and substantial in case of pearl millet. The sorghum production became somewhat drought resistant in the post-green revolution period and the production of pearl millet became highly drought sensitive.

Jha (1994), while conducting study on growth and instability in agriculture associated with new agricultural technology during 1972-73 to 1990-91, found declining instability in gross return and yield over years. The decline in yield instability in crop, viz., paddy and wheat was brought about with increased area under irrigation over the years. Nevertheless, the government's consistent price policy also helped in reduction of instability in farm harvest prices. Thus, it has been inferred that with new technology, instability in agricultural income reduced with adequate irrigation facilities and consistent price policy. However, the hypothesis of high instability in agriculture accompanied with high growth rate was established.

Singh and Mathur (1994), in their study "Growth and Instability in the Production and Price of Potato in India" found that the production of potato increased rapidly mainly due to the significantly high growth of area in the major growing states, yield also showed a rising trend but its growth was low in some states. Hence, there is a need to raise the yield by the use of improved seeds and package of practices as recommended by the extension organisations of the regions. Besides, the better infrastructural facilities such as irrigation water, marketing and storage should be provided to the potato growers in the identified states. Results further revealed that the production of potato was unstable in nature particularly in the states where the production increased at a fast pace due to adoption of modern production technology which led to instability in production. The area under potato was more unstable than yield in some of the states. This indicates that acreage of potato is responsive to comparative returns and prices of its competing crops in the region. Hence, variability in area may be reduced through suitable policies which ensure remunerative and stable prices to the potato growers.

Sawant and Achuthan (1995) in their study entitled "Agricultural Growth across Crops and Regions in the Post-Green Revolution Period" unfold interesting dimensions of the new trends and patterns emerged in the 1980s. The fact that the role played by yield improvement in inducing higher output growth has been far more important than that of expansion in area which indicates technologically induced process of growth in potato production.

In recent years, acceleration in yield growth has been significant for crops like rice, maize, other pulses, rapeseeds, seasamum, soyabean, rubber and cotton, however, the yields of other crops continued to expand at the rates close to their pre-1981 levels. Again, unlike the past, enhancement in yield growth has been more impressive for non-foodgrains as compared to foodgrains, and among the foodgrains for Kharif foodgrains vis-a-vis the rabi foodgrains, this is an indication of a much wider diffusion of technology across the crops. Yet another distinguishing character of agricultural growth in the 1980s has been its wider dispersal over the regions.

Factors Affecting Agricultural Production

A large number of research studies have been conducted on factors affecting agricultural production at national and state levels but most of them are based on tabular analysis and very few have used the reliable statistical tools. Besides, most of such type of studies are based on cross-sectional data collected from limited number of respondents from very localised area. In the present section, the studies based on time series data and directly related to the concerned objective of the present study are reviewed.

Giri *et al.* (1966) attempted to study the contribution of land, irrigation and fertilizer to the growth of crop output at national level. The analysis covered the period 1951-52 to 1962-63. He concluded that land continued to be the major contributor of crop output in India and that irrigation and fertilizer had yet to play their significant role in increasing agricultural production during the period under study. It may be worth mentioning that the study was conducted prior to green revolution.

Parikh (1966) in his study "State-wise Growth Rate in Agricultural Output: An Econometric Analysis", measured the contribution of land, irrigation and fertilizer to the growth of crop output by using the Cobb-Douglas model of functional analysis. He found that the coefficients for land were positive and significant in all the states, except Tamil Nadu. The growth of crop output was explained by land to the extent of 70 per cent or more. The coefficients for irrigation did not indicate uniform trend. Fertilizer has positive coefficients in all the states except for Uttar Pradesh.

Abraham and Raheja (1967) examined the growth of production of rice and wheat crops in India during the period 1951-52 to 1964-65 and analysed the contribution of inputs like land area, irrigation and fertilizers to the growth in rice and wheat output. They found that these three factors accounted for 7.16, 15.84 and 77 per cent of growth, respectively in rice production. In case of wheat production, these inputs shared about 30, 15 and 35 per cent, respectively.

Randhawa and Dev (1972) in their study on "Fertilizer Usage in Changing Agriculture in Punjab", found that fertilizer contributed in increasing agricultural production to the extent of 45 per cent, while other factors like irrigation, improved seeds, double cropping and land reclamation measures contributed only 27, 13, 10 and 5 per cent respectively. The further probe into relationship and contribution of fertilizer to wheat yield in nine districts of the state revealed that as much as 70-80 per cent increase in yields in the state ascribed to increased use of fertilizer alone.

Mukhopadhyaya (1976), while conducting a study on source of variation in agricultural productivity in India reported that inputs accounted for only 41 per cent of total variations in output in different wheat growing regions of the country. The remaining 59 per cent variation was due to factors like characteristics of soil, climate and temporal factors. And out of the latter 95 per cent variance of total distribution is attributed to the regional and 5 per cent to temporal cause. Ludhiana in Punjab stood the highest and Allahabad in Uttar Pradesh the lowest where regional factors' coefficients were 5.47 and 0.36 respectively. This means that if Ludhiana uses the same quantity of land, irrigation, fertilizers, tractors, literate and illiterate labours as Allahabad does, the output in Ludhiana is likely to be 15 times more than that in Allahabad.

Sagar (1978), while analysing contribution of individual technological factors in agricultural growth in Rajasthan during 1961 to 1974, found that the analysis contradicts the popular hypothesis that the new varieties are, by and large, responsible for the green revolution in India. Even in the case of wheat which registered nearly 40 per cent yield growth, the share of factors like fertilizers and irrigation was estimated to 74 per cent and the share of new varieties of seed was 15 per cent only. The share of fertilizers and irrigation was 30 per cent and 18 per cent, respectively to the growth of overall agricultural productivity, and 33 per cent variation was unexplained in the study.

Raj (1979) analysed the economic trends and changes in strategy of crop output in India. He agreed with the viewpoint that irrigation is the crucial factor in agricultural growth. He further observed that increase in agricultural output has been achieved in a few states, whereas area under irrigation had been extended the most. For instance, nearly three-fifth of increase in foodgrain output between 1967-68 and 1977-79 (from 95 to over 125.5 million tonnes) was in six states of Haryana, Tamil Nadu, Uttar Pradesh, Andhra Pradesh, West Bengal and Punjab. The conclusion is based on tabular analysis.

Ram (1979) analysed the contributions of area, yield and cropping pattern to total output in the eastern region of Uttar Pradesh. The analysis revealed that contribution of gross cropped area in the output growth in the entire region had been 23.66 per cent. The yield had contributed about 46.97 per cent to the growth of agricultural output in the region. In the whole of the Ganga-Ghaghra doab, except the districts of Faizabad and Sultanpur, the yield did not seem to have contributed to output growth significantly, while in the north-eastern parts of the region, the contribution of yield became more than 65 per cent. The share of cropping pattern to output accounted for 64.60 per cent which ranked first among all the components contributing to the output growth. It was further observed that the contribution of cropping pattern to the output growth was high in those areas where the contribution of yield was less and vice versa.

Dhawan (1991) conducted a study on "How Stable is Indian Irrigated Agriculture?" and observed that no single state can be described to have achieved drought-proofing of its agriculture through irrigation. The signs of drought-proofing area are visible in the states of Punjab, Rajasthan and Karnataka.

In respect of six out of 11 states, it has been observed that the irrigation could afford partial protection in the sense that the decline in irrigated output turns out to be less than the corresponding decline in unirrigated output as a result of drought. In the absence of irrigation, impact of a drought on agricultural output, especially on foodgrains, tends to be more in low rainfall regions. According to data, the natural magnitude of the output elasticity with respect to rainfall tends to be about 1.6 in the low rainfall states, 1 in the medium rainfall states, and about 0.6 in the high rainfall states. In the presence of irrigation, these very magnitudes become 0.2, 0.5 and 1 respectively. This suggests that protective benefits of irrigation, even when partial, tend to be confined primarily to low and medium rainfall areas. In high rainfall areas, irrigation appears to aggravate the adverse output impact of a drought.

Kumar and Kaul (1991), while conducting a study on "Estimation of Contribution of Growth Component and Instability in Agriculture", reported that the coefficient of gross cropped area was positive and significant in all the models with and without time variable. The coefficients of gross irrigated area and fertilizer consumption were not found to be statistically significant during period. The value of coefficient of multiple determination (R^2) was about 91 per cent in all the cases indicating that 91 per cent of the variation in the agricultural output was explained by variables like gross cropped area, gross irrigated area, quantity of fertilizer consumed per unit of irrigated area, and proportion of area under HYV to gross cropped area.

The contribution in growth of agricultural output by gross cropped area, irrigation, fertilizer and area under HYV were 72, 18, and 5 per cent respectively, during the period 1966-82 indicating that the land was an important factor for agricultural growth in early green revolution period.

A perusal of literature discussed under above three subheads did not provide a definite direction because these studies were conducted in different time periods and locations. The empirical studies on cropping pattern conducted in post-green revolution period were reviewed. It has been observed that the crops like, rice and wheat gained area in almost all the states during post-green revolution period. On the other hand, oilseeds, fruits, vegetables and non-food crops gained area in Jammu & Kashmir state, probably due to favourable agro-climatic situation of the state for growing these non-foodgrain crops. Cash crops like potato, onion, garlic and other spices could not find place in cropping pattern analysis, probably due to non-availability of data on these crops. Almost all the studies under review dealt with the cropping pattern scenario and changes at national and outmost to state level. There has been dearth of such type of studies dealing with regional scenario of cropping pattern.

A large number of studies have been conducted on growth and instability is agricultural production in general and foodgrain production in particular. Among the principal crops, wheat gained area, production and productivity in almost all the states during post-green revolution period. Rice also gained area, production and productivity, particularly in the states which adopted rice cultivation during post-green revolution period. Pulses and oilseed did not show any sustained growth in area across the states but productivity of these crops picked up in most of the states. There has not been any clear-cut trend in stability in production of principal crops. Hazell (1982) found rainfall, irrigation and agro-physical situation as important factors influencing instability in production of foodgrain crops during post-green revolution period. Moreover, the application of new technology has also been found responsible for absolute yield variability. The interplay of agro-climatic factors and new technology led to increase in production instability in India, however, HYVs were not found responsible for this (Pal and Sirohi, 1991).

Most of the studies compared the growth and stability performance of pre-and post-green revolution periods and almost all of them dealt either with national data or state data. If rainfall and agro-climatic situations are major factors influencing stability which varies much even within states hence regional studies dealing with different periods of post-green revolution period would generate much meaningful information on these aspects.

A few studies have been conducted to identify factors responsible for production and growth in foodgrain production, particularly by analysing time series data, though these were again conducted at national/state level. No effort seems to have been taken to analyse these factors at regional level. Hence, the present study is an effort to analyse these aspects at regional level during post-green revolution period because no such effort has been made in the past, particularly with respect to Bihar in general and North Bihar in particular.

3

Methodology

This chapter deals with the methods adopted to obtain and analyse data to test hypotheses of the present study. This presents appropriate research design, methods and tools used to obtain the evidences required and the analytical procedure followed to reach at the relevant conclusions to be highlighted in the subsequent sub-sections. The details of the procedures, thus, followed are presented in this chapter.

Selection of the Area

Agriculture still contributes nearly 50 per cent to the gross domestic product (GDP) and provides employment to more than 80 per cent of workforce in North Bihar. The recurrent flood menace in North Bihar makes the region quite agriculturally different from other regions of the country in general and South Bihar in particular. Keeping in view these peculiarities, North Bihar was selected for detailed investigation of agricultural growth and instability. There are two agro-climatic zones in North Bihar, i.e., Zone I and Zone II. The names of the districts of Zone I and Zone II are given in Table 3.1.

Table 3.1
Names of the Districts in Two Zones of North Bihar

Zone	*Districts*
Zone-I	Saran, Siwan, Gopalganj, Muzaffarpur, East Champaran, West Champaran, Sitamarhi, Vaishali, Darbhanga, Madhubani, Samastipur, Begusarai.
Zone-II	Saharsa, Purnea, Araria, Madhepura, Khagaria, Kishanganj, Katihar, Navgachia* subdivision of Bhagalpur district.

* Data of this subdivision were not included in the study due to non-availability of time series data.

Sources of Data

The study is based on secondary data. General information and district-wise time series data on area, production and productivity of principal crops were obtained from the offices of the Directorate of Statistics and Evaluation and Department of Agriculture, Government of Bihar. District-wise fertilizer consumption data were collected from the *Fertilizer Statistics*—a publication of the Fertilizer Association of India. *The Economic Survey*, *Agriculture Statistics at a Glance*, Ministry of Agriculture, Government of India, and *Statistical Handbook*, Government of Bihar were also consulted for obtaining desired data.

Period of the Study

The study was planned to analyse the post-green revolution scenario of agricultural production since the finding of any investigation dealing with agricultural growth of pre-green revolution period is hardly relevant in the present context and would have limited use in formulating strategies for agricultural development. In India, green revolution commenced in mid-sixties but its widespread impact on farmers' fields was observed after five years in the project area. Hence, the present study covered the period 1970-71 to 1993-94.

It has been realised that the green revolution lost steam and the agricultural growth slided down from 2.32 per cent during 1960-70 to 1.90 per cent during 1971-85.[1] In this backdrop, an attempt has also been made to study the agricultural growth for three post-green revolution periods, i.e., 1970-71 to 1979-80, 190-81 to 1989-90 and 1990-91 to 1993-94. There is, no doubt, some arbitration in choosing the cut-off points for dividing the periods but it was done to study the agricultural growth during different periods of green revolution which can be termed as Period I (1970-71 to 1979-80), Period II (1980-81 to 1989-90) and Period III (1990-91 to 1993-94). Abnormal years could have been excluded for estimating growth rates to avoid its influence on estimated growth rate but it was not done due to presence of five abnormal years during the period under study, i.e., 1977-78, 1979-80, 1982-83, 1987-88 and 1990-91. Moreover, the precise understanding of growth and fluctuations can only be possible by considering good, as well as bad years in the analysis. The growth rate and fluctuations were also estimated from moving period of five years length. A set of successive five years was constituted from the entire series by excluding the first year of the previous and including the immediate next year from the series. The information generated through this analysis would serve as a supplement to the growth analysis for the three distinct periods. The five year length of the moving period has been considered to even out the probable effects of weather cycle which is of five years in the state of Bihar.[2] Growth rate and fluctuation estimates for moving periods were also utilised to find out regression estimates between growth and instability.

The contribution of area, productivity and their interactions to the instability in production of concerned principal crops was estimated for the study period. Regression analysis between growth rates and instability in principal crops were estimated through their moving estimates for two periods, i.e., early green revolution period (1970-75 to 1979-84) and latter green revolution period (1980-85 to 1989-94). It was done purposively to have an idea about the extent of relationship between growth and instability of principal crops in two distinct periods of green revolution.

Analytical Framework

As mentioned earlier, analysis was made for three periods, however, moving averages of five years have also been estimated for growth rate and fluctuation study of different crops. Analysis was done for the whole of North Bihar and zone-wise analysis has also been undertaken to have comparative performance of zones of North Bihar.

The cropping pattern of the project area is studied at only three points of time since the study of cropping pattern of all the 24 years from 1970-71 to 1993-94 would not give additional information since there is no evidence of significant change in cropping pattern between consecutive years. To have the normal area of different crops for comparative study, triennium of three succeeding years at each of three points of time were estimated, i.e., 1972-73-74, 1982-83-84, and 1992-93-94 which are termed as the year 1974, 1984 and 1994 respectively in the text.

Crops Included

The study deals with the area, production and yield of principal crops grown in the project area. These crops are rice, wheat, maize, barley, *arhar*, gram, lentil, rape and mustard and sugarcane. Moreover, all major crop groups like foodgrains, cereals, pulses, and oilseeds were also considered for analysis. The crops included in the study taken together constituted more than 80 per cent of gross cropped area. Principal crops like rice, wheat, maize, lentil and rape and mustard were considered for regression estimates between growth and instability and for assessing the contribution of area, productivity and their interaction to the instability in their production.

Analytical Procedure

The present study was designed primarily to examine the changes in cropping pattern, changes in area, production and productivity, and instability in production of principal crops along with its relation with growth during period under study. An effort was also made to measure the contribution of different components to agricultural production. The components included for this analysis are gross cropped area, gross irrigated area, quantity of fertilizer consumed and area under HYVs. The components of growth in foodgrain production were also estimated by considering factors of production which emerged as statistically significant in the functional analysis. The statistical tools used for analysis of data for drawing relevant conclusions are presented in subsequent subsections.

Change in Cropping Pattern

'Cropping pattern' here has been defined as distribution of acreages expressed in percentages of total cropped area (Ramasubban, 1963).[3] Cropping pattern determines the output-mix in different regions. It has significant bearing on widening the geographical inequalities in income distribution. It was, therefore, considered worthwhile to measure the change in cropping patterns over a period of time and across the zones.

Based on the share of each crop to total cropped area, percentages were worked out and ranks were assigned to each crop according to their respective proportion to gross cropped area in the particular year. Crop with the largest proportion was assigned first rank the second largest assigned second rank and so on. The concordance coefficient (w) was found out between the rankings of the different crops in the year ending 1974, 1984 and 1994 for North Bihar and both the zones separately. Concordance coefficient was used by Ramasubban (1963) and Sridharan and Radhakrishnan (1978),[4] in analysing the changes in cropping pattern in their respective studies which was adopted for this analysis. The formula used for calculating the concordance coefficient (w) is given as below:

$$w = \frac{12 \sum_{i=1}^{n} (x_i - \bar{x})^2}{m^2 (n^3 - n)}$$

where,

$$x = \frac{m(n+1)}{2}$$

x_i = sum of the ranks allotted to ith crop over the period under study (i = 1,2, 3, ... n)

m = number of years

n = number of crops

The value of concordance coefficient measures is the degree of agreement between number of crop sets. A complete randomness in ranking leads to w=0 on the one hand, and perfect agreement among patterns result in w=1, on the other. The greater the departure from complete agreement, the smaller is the value of concordance coefficient. If 'w' takes only non-zero value, whatsoever, then there is certain amount of agreement among the cropping patterns. In the present study, Chi-square (X^2) will be calculated by using following formula:

$X^2 = M(n - 1)w$

If table value exceeds the calculated X^2 value,[5] the coefficient of concordance is not significant and the ranking in different cropping pattern for the crops are compatible.

Area, Production and Productivity Changes

Exponential Trend Equation: In order to find out the compound growth rates with respect to area, production and productivity of important crops, exponential function was used:

$$Y_t = AB^t \quad \text{... (1)}$$

where

$Y_t =$ area/production/productivity of crop concerned in the year

$A =$ intercept,

$t =$ year

$B =$ 1 + r/100

where, 'r' refers to the percentage rate of compound growth of area/production/productivity of concerned crop per annum.

By taking logarithm of both sides of the equations, we get:

$$\ln Y_t = \ln A + t \ln B \quad \text{... (2)}$$

If we put in A=a, and 1n B=b, then the above equation can be written as:

$$\ln Y_t = a + bt \quad \text{... (3)}$$

By using ordinary least square technique, we have normal equation of the type:

$$\ln Y = Na + bt \quad \text{... (4)}$$

$$1n\ Y_t = at + bt^2 \qquad \dots (5)$$

where, N is the number of observations (years).

By solving equation (5), the value of 'a' and 'b' were computed. For deriving compound growth rate from the regression coefficients, the following procedure was adopted.

When 'b' had positive value, antilog of b was obtained and then '1' was subtracted from antilog value of 'b'. Then b-1 value was multiplied by 100. This gave the compound growth rate of increasing type. When 'b' had negative value, the procedure of calculating compound growth rate was the same as it was in the case of 'b' having positive value. The negative compound growth rate indicated the decreasing rate over time.

Percentage rate of compound growth per annum was calculated as:

$$r = (B\text{-}1) \times 100 \text{ or } (\text{antilog } B\text{-}1) \times 100$$

which represents a rate of change from observation to observation during the period under study.

Moving period (5 years) compound growth rates in area, production and productivity of principal crops were also estimated for further analysis like regression estimates between growth and instability and for showing trends in area, production and productivity through graphs.

Measurement of Instability in Production

Purpose of this analysis was to study the fluctuations in output from its trend. Hence annual per cent changes in output and then the standard deviation in such annual changes were estimated which gave a measure of fluctuations around the trend line. The following formula was adopted for calculating standard deviations.

$$\text{Standard Deviation} = \sqrt{\frac{1}{N}\sum_{i=1}^{N}(x_i - x)^2}$$

where,

N = number of observations

x_i = value of ith observation

$$x = \frac{1}{N}\sum_{i=1}^{N} x_i = \text{mean of } x_i$$

Variability in annual output growth rates provides a measure of the degree of instability in production over the specified period. Moreover, decomposition of the variability in annual output growth rates may help to identify the sources of change in instability in production over different periods. Besides the institutional, technological, economic and weather factors, the quantifiable and most closely associated factors like area, productivity and the interaction are likely to influence instability in agricultural production. The variability in annual output growth rates over a specified period of length 'T' can be decomposed as:

$$V(Go) = V(Ga) + V(Gy) + 2\,Cov\,(GaGy)$$

where, Ga, Gy and Go are the growth rates in area, yield and output estimated over the period under study.

Measurement of Components Affecting Agricultural Production/Growth

For the measurement of the components to production/ growth, the following form of function was used:

$$Y' = B'_0 + B'_1 X_1 + B'_2 X_2 + B'_3 X_3 + B'_4 X_4 + B'_5 X_5 + u \quad \text{... (A)}$$

$$\text{Log } Y' = \text{Log } B'_0 + B'_1 \log X_1 + B'_2 \log X_2 + B'_3 \log X_3 + B'_4 \log X_4 + B'_5 \log x_5 + \log u \quad \text{... (B)}$$

where,

Y = annual production of foodgrains (in tonnes)

X_1 = gross cropped area (in ha)

X_2 = gross irrigated area (in ha)

X_3 = fertilizer consumption (in tonnes)

X_4 = HYVs area (in ha)

X_5 = annual rainfall (in mm)

Least square estimates of regression coefficient of B'_1, B'_2, B'_3, B'_4, and B'_5, were computed and step down procedure based on conventional t-test was adopted for identifying the variables, which had significant contribution towards the response variable 'Y'.

In order to study the relative importance of the variables which had a significant effect on Y (dependent variable), the regression sum of square was splitted into different orthogonal sum of squares, each based on one degree of freedom.[6] Following this procedure, the coefficient of determination (R^2) which was defined as the ratio of regression sum of square was partitioned into different components, each component signified the percentage variability in dependent variable (Y), explained by a significant variable in the regression model obtained from step-down procedure. The step-down regression analysis suggested a significant contribution of variables X_1 ,X_2, and X_5 towards Y then the model (B) reduced to:

$$\text{Log } Y = \log B_0 + B_1 \text{ long } X_1 + B_2 \log X_2 + B_5 \log X_5 \quad \text{... (C)}$$

The R^2 value is given by

$$R^2 = (B_1 x_1' y' + B_2 x_2' y' + B_5 x_5' y')/y'^2$$

$$= \frac{B_1 x'_1 y'}{y'^2} + \frac{B_2 x'_2 y'}{y'^2} + \frac{B_5 x'_5 y'}{y'^2}$$

$$= {}_{i=(1,2,3)}\Sigma R^2$$

where,

R_i (i =1,2,5) is the percentage variability explained by the variables X_1

$y' = Y - Y$

$x'1 = X_1 - X_1$

$x'2 = X_2 - X_2$

$x'5 = X_5 - X_5$

The coefficients of gross cropped area, gross irrigated area and rainfall were found positive and significant. Therefore, this model was selected for estimating the contribution of these variables towards the growth of agricultural output. The following model was used:[7]

$$Gy = B_1Gx_1 + B_2Gx_2 + B_5Gx_5$$

where, Gy, Gx_1, Gx_2 and Gx_5 are the growth rates of agricultural output, gross cropped area, gross irrigated area and annual rainfall, respectively and B_1, B_2 and B_5 were respective regression coefficients of above mentioned variables.

Limitations

A well-thought and designed research methodology may not be free from limitations, hence the present methodology adopted for analysing data in the present investigation may have inherent limitations. The main limitations of the present study are mentioned below:

1. While analysing the cropping pattern, principal crops are included but minor commercial crops like spices, fruits and vegetables have not been included due to unavailability of time series secondary data on these crops. Hence, the investigation failed to draw any conclusion on increasing area under these commercial crops, particularly during post-liberalisation period.
2. The conclusions drawn from the Period III, which encompasses only four years, may not hold true because the period is too small to provide any reliable information. Moreover, it will provide a direction in agricultural production and growth which can be used for formulating research proposals in future.
3. The result might have been different for rice and maize crops when season-wise area and production of these crops would have been taken for analysis because winter maize is quite different from kharif maize and summer rice is different from kharif rice.
4. Unlike other similar studies, the abnormal agricultural years have not been deleted for estimating growth and instability. It was done purposively to have the real trend in growth and instability of agricultural production because the deletion of abnormal years might have provided a better growth rate and more stable agriculture which was far from the real situation.

Notes

1. Hanumantha Rao, C.H., Ray, S.K. and Subbarao, K. (1988), *Instable Agriculture and Drought*, Vikas Publishing House Pvt. Ltd., New Delhi, p.13.
2. Sen, S.R. (1991), "Growth and Instability in Agriculture", *Recent Advances in Agricultural Statistics Research*, Wiley Eastern Limited, New Delhi, p. 3.
3. Ramasubban, T.A.(1963), "Some Statistical Measures to Determine Changes in Cropping Pattern", *Agricultural Situation in India*, 17(11), p. 1153.
4. Sridharan, B. and Radhakrishnan, A.S. (1978), "A Study on the Factors Affecting Changes in the Cropping Pattern in Nilgir District of Tamil Nadu", *Indian Journal of Agricultural Economics*, 33(4), p. 14.
5. Raghavarao, D. (1983), *Statistical Techniques in Agricultural and Biological Research*, Oxford and IBH Publishing Co., New Delhi, p. 333.
6. Myers, R.H. (1971), *Response Surface Methodology*, Allyn and Bacon, Inc., Boston, p. 48.
7. Kumar, Ashok and Kaul, Sushila (1991), "Estimation of Contribution of Growth Components and Instability in Agriculture", *Recent Advances in Agricultural Statistics Research*, Wiley Eastern Limited, New Delhi, p. 76.

4

Agro-Economic Background of the Project Area

Before going with the analytical description of the project, it is considered worth while to have acquaintance with the agro-climatic situation and socio-economic milieu of the project area, i.e., North Bihar. Thus, the pertinent information regarding the agro-economic situation of the project area has been presented in this chapter which includes location, soil and climate, demographic features, occupational distribution, degree of fragmentation, land utilisation pattern, area under different sources of irrigation, fertilizer consumption, cropping intensity, agricultural loans and infrastructural facilities, etc., of the project area.

Location

Bihar, one of the states of eastern India, is located between 21°58'10" to 27°31'15" north latitudes and 83°9'50" to 88°17'47" east longitudes and covers an area of 1.74 lakh sq km surrounded by four states and an international border, i.e., Uttar Pradesh and Madhya Pradesh in the west, Orissa in the south-west, West Bengal in the east and Nepal in the north. The state is divided into three physiographic regions, i.e., North Bihar, South Bihar plains, and Chotanagpur and Santhal Paragana plateau.

North Bihar with an area of 0.53 lakh sq km is located between 25°10' to 28°31 north latitudes and 83°45' to 88°17' east longitudes. This tract is surrounded by Nepal in the north, West Bengal in the east, the Ganga river in the south and Uttar Pradesh in the west.

North Bihar has been divided into two agro-climatic NARP zones, i.e., Zone I and Zone II. Zone I comprises districts of West Champaran, East Champaran, Gopalganj, Siwan, Saran, Sitamarhi, Muzaffarpur, Vaishali, Madhubani, Darbhanga, Samastipur and Begusarai. Zone II comprises districts of Khagaria, Saharsa, Madhepura, Purnea, Kishanganj, Araria, Katihar and Naugachhia subdivision of Bhagalpur district. The region is endowed with abundant groundwater and favourable agro-climatic situation for crop production.

Topography and River Basin

The topography of Zone I is practically plain having slope towards south-east. The flat alluvial plain has average altitude varying between 31 to 61 m from sea level with the exception of small range of hills, namely, Dun and Someshwar hills. There are three major river basins in this zone, namely, Gandak, Adhwara and Kosi. Due to silt inflow into these rivers, their beds have risen resulting in recurrent floods and deposition of sand silts all over the area. The depressions caused due to change in the course of these rivers remain submerged in water during rainy season and a vast area of this zone is under *chour* or *manes*.

Zone II, comprising the alluvial plains of Kosi, Mahananda and Ganga rivers, is slightly undulating to rolling landscape with a general slope towards south-east intermingled with long stretches of nearly flat landscape with pockets having subnormal depression where water accumulates. The area is full with braided streams (abounded beds of Kosi) and dead channels of Kosi and its tributaries. Small lakes, oxbow lakes and patches of marshy ground area are frequently encountered which might have been formed due to sudden and frequent changes in the courses of rivers Kosi and Mahananda. In southern fringes of the Zone II, between the natural levee of the Ganga on the one hand, and Kosi and Mahananda on the other and further within the embankments of Kosi, there are vast tracts similar to Tal land of South Bihar which become inundated and remain water-logged for a considerable period of the year. The general landslope is along south-east. The two major rivers Kosi and Mahananda with their tributaries which originate from the Himalayan region and merge with the Ganga have separate flood plains. However, flooding and sediment action pattern of these two rivers are somewhat different. Kosi is the longer river and has wider flood plains as compared to Mahananda.

Topographical Features Affecting Climate/Rainfall

In general, the topographical features of North Bihar plain do not influence the climatic conditions of this region. However, North Bihar remains humid due to presence of a large number of water bodies. The temperature conditions of North Bihar in general are not determined by its topography but cold and heat waves through convection have marked influence on the temperature of this region. The north-east portion of the North Bihar receives comparatively more rainfall. This may be due to geographic condition and/or by virtue of being situated in north of normal track of monsoon trough.

Soil and Climate

The soil of North Bihar is calcareous and the calcium percentage ranges between 5 and 40 per cent. The districts where this type of soil is found are East and West Champaran, Gopalganj, Saran, Siwan, Vaishali, Muzaffarpur, southern part of Darbhanga, Samastipur and Begusarai. Moreover, the soil of Siwan and Gopalganj is comparatively more alkaline.

The regions around Kosi and Mahananda rivers are endowed with non-calcareous soils. The districts with this type of soil are Sitamarhi, Madhubani, north portion of Darbhanga, Saharsa, Purnea, Katihar, Bhagalpur and Khagaria. The soil of north-east part of Purnea and the adjacent part of Nepal is light acidic.

The climate of study area is subtropical monsoon type. The climate of North Bihar is diverse with persistently lower range of temperature than that of south plains. The most important climatic factor affecting agriculture is rainfall. The average rainfall of North Bihar is fairly high, ranging from 1,050 mm to 1,700 mm, however, its distribution is uneven. A considerable portion of the rainfall is concentrated between the months of June and September. The North Bihar experiences too hot in summer season on the one side, and on the other, the temperature is too low in winter season. The months of April and May are dry and the hottest period when the temperature is about 37°C. December and January are the coolest months when the minimum temperature falls as below as 4°C. The average daily sunshine duration ranges between 5 and 7; 7 and 9; and 8 and 10 hours during kharif, rabi and summer seasons, respectively.

Abnormal Weather Phenomena

Low temperature accompanied with cold waves is commonly

experienced from 48th to 5th meteorological weeks (December and January). During this period, foggy weather continues for several days resulting in poor growth of rabi crops due to reduced availability of solar radiation. The westernly dry wind with high velocity usually starts blowing after 15th of March which causes forced maturity of late sown rabi crops, particularly of wheat thereby reducing crop yields. Unseasonal and Prolonged rains during March-April cause damage to mature crops standing in the fields and also to the harvested produced kept on the threshing floor. Further these rains reduce the output of thresher by increasing relative humidity. Isolated cyclonic and conventional rains accompanied with hailstorms during March-April cause tremendous damage to standing rabi crops through mechanical destruction of plants and/or harvestable portion of the crops. The frequency of storms is higher between April and June which results in severe damage to fruit crops.

Traditional Cropping Seasons

There are three traditional growing seasons, namely, kharif (rainy season), rabi (winter season) and zaid (summer season). Kharif season sets in with the onset of monsoon rains in North Bihar in the second week of June and continues till the end of September. This is the most important growing season and has the highest net sown area under rainfed crops like rice, maize, and pigeon pea. Rabi season commences from October-November and continues upto February-March. During this season, traditional mixed cropping of cereals with oilseeds and pulses is in vague. These crops are grown on residual moisture. In summer season (March to June) short duration crops like seasame, green gram and millets are mainly grown.

Demographic Features

Population of the Study Area

North Bihar with 30.8 per cent of geographical area supports 47.05 per cent of population of the state of Bihar, however, nearly 72 per cent of population lives in Zone I and 28 per cent in Zone II (Table 4.1).

Density of population in the study area is 727 persons/sq km which is comparatively higher than the density of the state (497 persons/sq km). At the zonal level, population density is comparatively higher in Zone I (846 persons/sq km) than in Zone II (609 persons/sq km).

Table 4.1

Demographic Features of the Study Area (1991)

Zone	*Population (in million)*	*Density (persons/ sq km)*	*Rural Population (%)*	*Schedule Caste (%)*	*Schedule Tribes (%)*	*Female literacy (%)*	*Average Land Holding (ha)*	*Rural Poverty (%)*
I	29.19	846	93.03	14.20	0.28	15.2	0.62	49.7
II	11.45	609	92.06	17.18	2.38	12.19	0.84	49.7
North Bihar	40.64	727	92.54	15.99	1.33	13.69	0.73	49.7
Bihar	86.37	497	86.86	14.55	7.66	23.89	0.90	51.0

Source: Population Census 1991 and *Bihar ek Jhalak*, Directorate of Statistics and Evaluation, Government of Bihar, Patna, 1992.

Proportion of scheduled caste (SC) population to total population of the study area is comparatively higher (15.99%) than the corresponding proportion for the state (14.55%). The proportion of scheduled tribes (STs) is only (1.33%) of the total population in the study area. Of the two zones under study,

Zone II has comparatively higher proportion of SC population (17.78%) than Zone I (14.20%). The Zone II has 2.38 per cent of tribal population whereas the proportions of tribal population to total population in Zone I is 0.28.

The female literacy is poor in the project area (13.69%) which is nearly 10.2 per cent lower than the female literacy in the state (23.89%).

There has been rural-based economy in the North Bihar. In the study area, nearly 92.5 per cent of population lives in rural area as against 86.86 per cent in the state of Bihar, and 74.3 per cent in the country. The proportion of rural population is comparatively higher in Zone I (93.03%) than Zone II (92.06%).

Occupation Distribution

Occupational distribution of the North Bihar has been presented in Table 4.2.

Main Agricultural Workers

The majority of rural workers are engaged in agriculture and allied activities who are categorised as agricultural workers. The proportion of agricultural workers to total main workers of North Bihar varies from 89.01 per cent in Zone I and 93.33 per cent in Zone II. The corresponding proportion for the state is 81.07 per cent.

Agricultural Labourers

Agricultural labourers are workers who earn their livelihood mainly by working on others' farms and majority of them are landless labourers. Number of agriculture labourers is 5.13 million in North Bihar, accounting for nearly 46.64 per cent of rural main workers. The corresponding proportion is 37.12 per cent in the state. The proportion of agricultural labourers to rural workers is comparatively higher in Zone II (52.17%) than Zone I (44.11%).

Table 4.2

Occupational Distribution in North Bihar (1991)

Particulars	*Zone I*	*Zone II*	*North Bihar (Project Area)*	*Bihar*
Rural Population (million)	27.15	10.54	37.69	75.02
Main Workers (million)	7.55	3.45	11.00	25.62
Rural Population (%)	27.81	32.73	29.19	34.15
Agricultural Workers (million)	6.72	3.22	9.94	20.77
Main Workers (%)	89.01	93.33	90.36	81.07
Agricultural Labourers (million)	3.33	1.80	5.13	9.51
Main Workers (%)	44.11	52.17	46.64	37.12
Non-farm Workers (million)	0.83	0.23	1.06	4.85
Main Workers (%)	10.89	6.67	9.64	18.93
Agricultural Wage Paid (Rs)	15.00	16.00	15.50	NA
Statutory Agril. Wage (Rs)	27.00	27.00	27.00	27.00

Source: Census 1991, Government of India.

Agricultural Wages

The statutory agricultural wage has been fixed at Rs 27 for eight hours of work but it has not been enforced strictly in the state. The main reason for non-payment of statutory wage in the state is low productivity of agricultural labour.

Opportunity for Non-Farm Employment

There are 1.06 million non-farm workers in North Bihar as against 4.85 million in the state. The proportion of non-farm workers to rural main workers is 9.64 per cent in North Bihar and 18.93 per cent in the state, indicating comparatively less non-farm employment in North Bihar in general and Zone II (6.67%) in particular.

Degree of Consolidation/Fragmentation of Operational Holdings

An increase in the number of operational holdings during the last two decades clearly indicates a greater degree of fragmentation resulting in a large number of uneconomic size of holdings in the study area. An average number of fragments per household and per hectare of land were estimated to five and four in Zone I and seven and three in Zone II respectively (Table 4.3). It is worth pointing out that the fragments are situated at an average distance of 0.10 to 2 km.

The consolidation of holdings is, no doubt, one of the important components of land reform but needs institutional efforts and political will. In North Bihar, the work on consolidation of holdings was taken up in early sixties but it could not succeed.

Effect of Fragmentation/Consolidation

It is a common knowledge that the fragmentation of holdings affects the farming efficiency adversely, whereas the consolidation of holdings is considered to be most effective in improving the farming efficiency.

Table 4.3
Average Operated Area, Number of Fragments and Distance of Fragments from Households

		Operated Area (ha)	*No. of Fragments*	*Distance from Households (km)*
A. Rainfed				
(i)	Small			
	Zone I	0.89	6 (1-12)	0.25 (0.25 to 0.75)
	Zone II	1.60	5	0.25 (0.10 to 1)
(ii)	Large			
	Zone I	3.7	3 (2-4)	0.5 (0.25 to 1.5)
	Zone II	5.2	15(4-60)	1.0 (0.25 to 1.5)
(iii)	Average			
	Zone I	1.45	5 (1-12)	0.50 (0.25 to 1.5)
	Zone II	2.32	7 (1-60)	0.30 (0.10 to 1.5)
B. Partially Irrigated				
	Zone I	6.2	11 (2-30)	2.0 (1 to 3)
	Zone II	10.8	3 (1-5)	1.0 (0.10 to 1.5)

Figures in parentheses indicate the range.

Source: A Report on Rainfed Farming in Bihar Plains, RAU, Pusa (1993).

In North Bihar, there is a marked difference in topography of agricultural land, viz., low land, medium land and upland. Majority of farmers of rainfed situation are subsistence farmers and grow a variety of crops under different land situations to meet their domestic requirements. In this endeavour, they prefer to have fragments in different land situations to grow wide variety of crops as per their domestic needs. The consolidation of land would deprive them of such practices.

Inspite of the above fact, some big farmers prefer to have

consolidated holdings and even under rainfed situation for better management.

Adoption of New Crops/Production Technologies

In rainfed upland situation, maize and millets like, kodo millet, fox-tail millet and barnyard millet were being grown traditionally. In recent years, upland rice has replaced kharif maize to some extent. Majority of the farmers are not aware of new technologies and improved crop varieties recommended for rainfed situation. However, farmers having partial irrigated holdings have adopted banana and boro (winter) rice in Zone II, rajmah in Zone I and Zone II. Introduction of winter maize in both the zones of North Bihar is most conspicuous.

Big farmers have started plantation of agro-forestry and horticultural crops in lands located at distant places. The area under mango and litchi orchards has increased considerably, particularly in Zone I.

Land Utilisation Pattern

Present section deals with a comparative analysis of land utilisation pattern of North Bihar, vis-a-vis, state of Bihar. The comparison is made at two points of time, i.e., pre-planning period (1950-51) and present situation (1993-94). Area of land put to different uses along with their proportions to respective geographical area are computed and presented in Table 4.4.

It may be observed from the table that the geographical area of North Bihar is 5.34 million ha, accounting for nearly 31 per cent of geographical area of the state of Bihar (17.3 m ha). Among the different uses of land, net area sown occupied comparatively larger proportion of land in both the situations, that is Bihar and North Bihar, which vary from 68.34 per cent to

Table 4.4

*Land Utilisation Pattern in North Bihar and Bihar at Two Points of Time**

Class of Landuse Pattern	*Pre-Planning Period 1949-1950-1951*		*Present Situation 1992-1993-1994*	
	North Bihar	*Bihar*	*North Bihar*	*Bihar*
Geographical Area	5.49 (100.00)	18.12 (100.00)	5.33 (100.00)	17.33 (100.00)
Forest	0.04 (0.70)	2.92 (16.16)	0.09 (1.79)	2.92 (16.87)
Barren and Unculturable Land	0.17 (3.15)	0.66 (3.67)	0.22 (4.17)	1.01 (5.86)
Land Put to Non-agricultural Uses	0.44 (8.13)	1.44 (7.99)	0.91 (17.23)	2.06 (11.92)
Culturable Waste Land	0.17 (3.19)	0.97 (5.37)	0.02 (0.54)	0.39 (2.27)
Permanent Pastures and Other Grazing Land	1.10 (2.00)	0.33 (1.82)	0.01 (0.25)	0.13 (0.76)
Miscellaneous Tree Crops and Groves	0.06 (1.21)	0.10 (0.57)	0.17 (3.32)	0.26 (1.54)
Other Fallow Lands	0.18 (3.31)	0.72 (4.01)	0.10 (1.98)	1.08 (6.24)
Current Fallows	0.54 (9.87)	1.86 (10.28)	0.40 (7.59)	1.84 (10.66)
Net Area Sown	3.75 (68.34)	9.08 (50.13)	3.36 (63.06)	7.60 (43.87)
Area Sown More Than Once	0.77 (14.93)	1.97 (10.91)	1.66 (31.12)	2.81 (16.23)

Figures in parentheses indicate the proportion of geographical area.
*Data relate to triennium ending 1951 and 1994.

Source: Directorate of Statistics and Evaluation, Government of Bihar.

63.06 per cent of geographical area in pre planning period and present situation, respectively in North Bihar and 50.13 per cent to 43.87 per cent of geographical area in the state of Bihar, respectively.

In Bihar, area under forest remained almost stagnant around 2.92 million ha during 1951-94. The proportion of forest area to geographical area increased marginally from 16.16 per cent in pre-planning period to 16.87 in present situation. On the other hand, area under forest in North Bihar was only 0.04 million ha in pre-planning period which increased to 0.09 million ha in present situation.

In North Bihar, barren and unculturable land and land put to non-agricultural uses increased continuously during study period, i.e., from 0.17 million ha and 0.45 million ha in pre-planning period to 0.22 million ha and 0.92 million ha in present situation respectively.

Area under culturable waste land and permanent pasturer and other grazing land declined continuously from 0.18 million ha and 0.11 million ha in pre-planning period to 0.03 million ha and 0.01 million ha in present situation respectively in North Bihar and from 0.97 million ha and 0.33 million ha in pre-planning period to 0.39 million ha and 0.13 million ha in present situation respectively in Bihar.

Miscellaneous tree crops and groves also increased continuously in both the situations, i.e., 0.07 million ha and 0.11 million ha in pre-planning period to 0.18 million ha and 0.27 million ha during the present situation in North Bihar and Bihar respectively.

Other fallow lands showed increasing trend in the state of Bihar which increased from 0.73 million ha in pre-planning period to 1.10 million ha in present situation whereas it declined continuously in North Bihar from 0.18 million ha in pre-planning period to 0.10 million ha in present situation. The current fallows occupied 7.59 per cent to 9.89 per cent of geographical area in North Bihar and 10.28 per cent to 10.66 per cent of geographical area in the state of Bihar, during the period. There has not been any specific trend in area under current fallow neither in the state of Bihar nor in North Bihar.

Irrigation

Table 4.5 reveals that the main sources of irrigation in North Bihar are wells, tubewells, and canals which command nearly 30 and 52.3 per cent of gross irrigated area in Zone I and 43.6 and 37.2 per cent of gross irrigated area of Zone II respectively during the year 1980-81, whereas in the year 1990-91 these sources accounted for 16 and 64.8 per cent of gross irrigated area of

Table 4.5

Source-wise Gross Area Irrigated in North Bihar

(area in 000 ha)

Source	*Zone-I*		*Zone-II*	
	1980-81	*1990-91*	*1980-81*	*1990-91*
Canal	239	179	149	181
	(30.0)	(16.0)	(43.6)	(31.8)
Wells and Tubewells	417	723	93	329
	(52.3)	(64.8)	(37.2)	(57.7)
Tanks	36	48	8	8
	((4.5)	(4.3)	(3.2)	(1.4)
Other Sources	105	166	40	52
	(13.2)	(14.9)	(16.0)	(9.1)

Figures in parentheses indicate per cent of gross irrigated area by all sources in respective zones.

Source: Directorate of Statistics and Evaluation, Government of Bihar.

Zone I and 31.8 and 57.7 per cent of gross irrigated area of Zone II respectively. Thus, wells and tubewells substituted the tanks, canals and other sources of irrigation during the period.

Area Under High Yielding Varieties

Area under HYVs for different crops have been presented in Table 4.6.

A perusal of Table 4.6 indicates that the percentage area under HYVs was found to be the highest in wheat followed by maize and rice. The area under HYVs of rice increased from 21.19 per cent in 1984-85 to 26.13 per cent in 1989-90 whereas the area under HYVs wheat rose to 78.35 per cent in 1989-90 from 78.01 per cent in 1984-85. But the area under HYVs maize declined to 70 per cent in 1989-90 from 74.92 per cent in 1984-85.

Table 4.6

Area Under High Yielding Varieties in the North Bihar

Crop	*1984-85*			*1989-90*		
	Total Area	*HYVs Area*	*Percentage of HYVs to Total Area*	*Total Area*	*HYVs Area*	*Percentage of HYVs to Total Area*
Paddy	2027.21	429.48	21.19	2210.06	577.51	26.13
Wheat	1021.95	797.26	78.01	1148.59	899.86	78.35
Maize	395.65	296.41	74.92	488.41	345.65	70.77
	3444.81	1523.15	44.22	3847.06	1823.02	47.39

Source: Directorate of Statistics and Evaluation, Government of Bihar, Patna.

Fertilizer Consumption

The fertilizer consumption in North Bihar was 2.96 lakh tonnes in the year 1987-88 which substantially rose to 3.24 lakh tonnes in the year 1991-92 (Table 4.7). The consumption of nitrogenous, phosphatic and potassic fertilizers increased from 1.93 lakh, 0.71 lakh and 0.32 lakh tonnes in the year 1987-88 to 2.09, 0.80 and 0.35 lakh tonnes in the year 1991-92 respectively.

Table 4.7

Fertilizer Consumption in North Bihar

(in lakh tonnes)

Type of Fertilizer	*1987-88*	*1991-92*	*Percentage Increased*
Nitrogenous fertilizer (N)	1.93	2.09	8.40
Phosphatic fertilizer (P_2O_5)	0.71	0.80	12.98
Potassic fertilizer (K_2O)	0.32	0.35	9.07
Total fertilizer (N+ P + K)	2.96	3.24	9.45

Source: Fertilizer Statistics 1988-89 and 1992-93 and Planning Commission.

Cropping Intensity

Gross cropped area, net area sown and cropping intensity at different points of time in North Bihar have been calculated and presented in Table 4.8.

Table 4.8

Gross Cropped Area, Net Area Sown and Cropping Intensity of North Bihar at Different Points of Time

Year	*Gross Cropped Area (in lakh ha)*	*Net Sown Area (in lakh ha)*	*Cropping Intensity*
1949-50-51	45.29	37.50	120.77
1963-64-65	45.08	35.47	127.11
1978-79-80	47.50	35.64	133.26
1988-89-90	48.70	33.65	144.71

Source: Directorate of Statistics and Evaluation, Government of Bihar, Patna.

It may be observed from the table that there has been a slight decrease in gross cropped area from 45.30 lakh ha at the triennium ending 1951 to 45.09 lakh ha at the triennium ending 1965 but there after it increased, though, the increase was more pronounced in the triennia ending 1980 (i.e., 47.50 lakh ha) but it again declined to 48.70 lakh ha at triennium ending 1990. Net area sown has also declined throughout the period understudy, except during the pre-green revolution period (1963-64-65). The trend was maintained in the early period of post-green revolution (1978-79-80) but fell down to 33.65 lakh ha in the present situation (1988-89-90).

Banking Scene

There are three institutional agencies operating in rural areas of North Bihar. Cooperative is the oldest financing organisation with two open bodies at the state level, i.e., Bihar State Cooperative Bank and Bihar State Land Development Bank. There were 3851 rural branches of scheduled commercial banks in Bihar at the end of March 1992, including 1995 branches of Regional Rural Banks and 1956 branches of commercial banks. The rural population per branch was estimated to be nearly 20,000 as against a norm of 10,000 population by National Commission on Agriculture (1976). The credit-deposit ratio has never crossed 40 per cent which has been much lower than national average of 60 per cent.

An agricultural loan of Rs 447 million was disbursed in the study area through institutional agencies in the year 1991-92 which includes Rs 212 million as short-term loan and Rs 235 million as medium and long-term loans.

Agricultural loan per hectare of gross cropped area was estimated to be Rs 91.78 in the North Bihar. It was, however, comparatively higher in Zone I (Rs 92.59) than Zone II (Rs

77.49). The amount of agricultural loan/ha in project area was much lower than the national average (Rs 215/ha).

Table 4.9

Zone-wise Short, Medium and Long-Term Agricultural Loans along with Loan per Hectare Gross Cropped Area (1991-92)

(in million rupees)

Types of Loan	*Zone-I*	*Zone-II*	*North Bihar*
Short-term Cooperative Loan	158	54	212
Medium-term + Long-term Loan	152	83	235
Total Loan	310	137	447
Loan per Hectare of Gross Cropped Area	92.59	77.47	91.78
Short-term Loan	47.09	30.41	43.52
Medium and Long-term Loan	45.50	47.06	48.25

Source: A Report on Rainfed Farming in Bihar Plain, RAU, Pusa (1993).

Infrastructure Development

Rural Roads

As many as 8794 villages, 56.38 per cent of total villages in North Bihar are connected with all weather roads (Table 4.10). About 59 per cent villages in Zone I and 50.1 per cent of villages in Zone II are connected by all weather roads. The comparatively higher proportion (71.5%) of villages of category 'A' are connected with all weather roads and the proportion declined with decrease in the size of villages under categories 'B', 'C' and 'D'. Further, zone-wise analysis indicates almost similar trend. A large proportion of bigger size villages is connected with all weather roads in Zone I (75.9%) and fallowed by Zone II (65.6%).

Table 4.10

Number and Percentage of Villages (in each size class) Connected by all Weather Roads in Different Zones of North Bihar as on March, 1992

Zone	*Villages by Size Class (Based on Population of the Village)*				*Total*
	A (1500 & above)	*B (1000-1500)*	*C (500-1000)*	*D (500 and below)*	
I	2007	703	1938	1802	6450
	(75.9)	(57.8)	(60.6)	(46.5)	(59.0)
II	686	343	572	743	2344
	(65.6)	(55.3)	(50.6)	(29.6)	(50.1)
North Bihar	2693	1046	2510	2445	8794
	(71.5)	(56.9)	(57.0)	(38.3)	(56.3)
State (Bihar)	4667	2663	6029	7614	20973
	(75.4)	(64.3)	(56.29)	(34.6)	(48.7)

Source: Annual Report of the Department of Rural Development, Government of Bihar (1992).

Rural Electrification

There are 16,828 electrified villages in North Bihar accounting for 71.76 per cent of total villages (Table 4.11). The proportion of electrified villages is, however, higher in Zone I (76.1%) than that of Zone II (60.5%). The number of electrified villages in the study area is, thus, higher than the state average (58.7%). Number of electrified villages accounting to size class are not available. In the study area, as many as 44 thousand of private and state tubewells and irrigation pumps have been energised upto March 1993, accounting for nearly 16.8 per cent of energised irrigation sources. The largest number of states energised irrigation pumps are in Zone I (13.9%) than that of Zone II (2.9%).

Table 4.11

Zone-wise Rural Electrification in North Bihar (as on 31.3.93)

Zone	*No. of Villages*	*Electrified Villages*		*Energized Irrigation Pumps*			*Total*
	(Including Uninhabi ted)	*Total*	*Harijan Basti*	*Private Tube-wells*	*State Tube-wells*	*Lift Irrigation Pumps*	
I	16925	12877 (76.1)	7387	33717 (13.2)	2235 (43.2)	426 (28.6)	36378 (13.9)
II	6527	3951 (60.5)	2595	7182 (2.8)	292 (5.7)	147 (9.9)	7622 (2.9)
North Bihar	23452	16828 (71.76)	9982	40899 (16.0)	2528 (48.9)	573 (38.5)	44000 (16.8)
State (Bihar)	79397	46648 (58.7)	21487 (100.0)	254878 (100.0)	5174 (100.0)	1488 (100.0)	261540 (100.0)

1. Figures in parentheses in column 3 indicate percentage of the villages electrified in the respective zone.
2. Figures in parentheses in column 5, 6, 7 and 8 indicate percentage of total number of respective type electrified irrigation structures of the state.

Source: Annual Report of Bihar State Electricity Board.

Among different categories of energised irrigation pumps, private tubewells constitute 97.40 per cent, state tubewells 2 per cent and lift irrigation 0.60 per cent of the total energised pumps in the state.

On the basis of above discussion, it may be mentioned that the North Bihar is distinctly different from Central Bihar and plateau regions, particularly with respect to agro-economic characteristics. Hence, any research study conducted for whole of the state may not be true for North Bihar. Hence, the findings of the present study should be applicable in situations like North Bihar.

5

Cropping Pattern and Agricultural Growth

Cropping pattern connotes the crop-mix grown in a particular area in an agricultural year. Introduction of new agricultural technologies has influenced the crop-mix which is more prominent in agriculturally developed states (Singh, 1980). During post-green revolution period, a decline in area under rice, *bajra*, *ragi*, barley and *arhar*, and an increase in area under wheat, maize and groundnut was observed in the state of Bihar (Mohammad and Singh, 1989). Keeping in view the dynamics of cropping pattern in the process of agricultural development, an analysis of cropping pattern of North Bihar has been undertaken which is presented in the present section.

The cropping pattern has been studied at three points of time, i.e., triennium ending 1974, 1984 and 1994. A comparative study of cropping pattern of Zone I and Zone II has also been pursued at three specified points of time.

Table 5.1 depicts the share of major crop categories in gross cropped area at three points of time. Perusal of the table indicated cereal dominated cropping pattern in North Bihar since

Table 5.1
Zone-wise Share of Major Crop Categories in Cropping Pattern of North Bihar at Three Different Points of Time (During Post-Green Revolution Period)

(in per cent)

Crop Categories	*Year Ending 1974*	*Year Ending 1984*	*Year Ending 1994*
North Bihar			
Total Cereals	74.50	74.97	79.63
Total Pulses	9.70	9.68	9.06
Total Oilseeds	1.83	1.92	2.09
Other Crops	13.97	13.43	9.22
Gross Cropped Area	100.00	100.00	100.00
Zone I			
Total Cereals	77.70	79.49	83.68
Total Pulses	9.50	8.87	8.66
Total Oilseeds	1.60	1.65	2.07
Other Crops	11.20	9.99	5.59
Gross Cropped Area	100.00	100.00	100.00
Zone II			
Total Cereals	67.42	65.43	71.18
Total Pulses	10.13	11.39	9.85
Total Oilseeds	2.34	2.49	2.14
Other Crops	20.11	20.69	16.83
Gross Cropped Area	100.00	100.00	100.00

nearly three-fourth of gross cropped area was covered by cereals which increased from 74.50 per cent in 1974 to 74.97 per cent in 1984 and 77.53 per cent in 1994. The Zone I had comparatively more cereal dominated cropping pattern since the comparatively higher proportion of gross cropped area (77.70 to 83.88%) was covered by cereals in Zone I than that of corresponding proportion in Zone II (65.42 to 71.18%). It is also worth pointing out that the proportion of cereals in cropping pattern increased in Zone I but declined in Zone II during the period under study.

In North Bihar, the share of pulses in gross cropped area declined from 9.70 per cent in 1974 to 9.68 per cent in 1984 which further declined to 9.06 per cent in 1994. The similar trend was observed in Zone I whereas in Zone II the pulses' share in gross cropped area increased from 10.13 per cent in 1974 to 11.39 per cent in 1984 but declined to 9.90 per cent in 1994. The comparatively higher proportion of gross cropped area was covered by pulses in Zone II (9.83 to 11.39%) than proportion of gross cropped area covered by pulses in Zone I (8.66 to 9.50%). Oilseeds, no doubt, is a crop of minor importance in the project area but the proportion of area under oilseeds to gross cropped area increased from 1.83 per cent in 1974 to 1.92 per cent in 1984 and 2.09 per cent in 1994. The similar trend has been observed in Zone I. In Zone II, the proportion of area under oilseeds to gross cropped area increased from 2.34 per cent in 1974 to 2.49 per cent in 1984, but marginally declined to 2.14 per cent in 1994.

Despite the agricultural development, cereals dominated in the cropping pattern of North Bihar which was more pronounced during the early part of 1990s. Oilseeds crops are, no doubt, unimportant crops with respect to area but gained area, during the period under study. On the other hand, pulses are important non-cereal crops but witnessed declining share in gross cropped area in both the zones during the period under study.

An attempt has also been made to study the cropping pattern with respect to proportion of principal crops in gross cropped area during period under study. Zone-wise cropping pattern constituting share of principal crops in respective gross cropped area at three specified points of time are presented in Table 5.2.

It may be observed from the table that the percentage area under major cereals like, rice and wheat increased from 42.42 per cent and 15.51 per cent in 1974 to 43.53 per cent and 18.92 per cent in 1984 respectively but the former declined to 42.01 per cent and latter increased to 23.35 per cent in the year 1994. The similar trend was observed in both the zones under study but the decline in the proportion of rice was more pronounced in Zone II during latter period of green revolution (1984-95).

Wheat emerged as an important cereal crop in North Bihar which occupied nearly one-fourth of gross cropped area (23.35%) in the year 1994. The increasing share of wheat in the gross cropped area was more pronounced in Zone I, since it occupied more than one fourth of gross cropped area (26.22%) and more than 50 per cent of gross cropped area covered in rabi season in the year 1994. Maize was the third important cereal in the cropping pattern in the project area. In North Bihar, the share of maize in cropping pattern remained almost identical during 1974-84, but showed declining trend from 10.04 per cent in 1984 and 9.26 per cent in 1994. A zone-wise analysis revealed that the share of maize in gross cropped area in Zone I declined continuously from 11.77 per cent in 1974 to 9.71 per cent in 1994. But the situation was just reverse in case of Zone II whereas proportion of maize in gross cropped area increased from 6.09 percent in 1974 to 7.98 per cent in 1984 and 8.38 per cent in 1994. The decline in proportion of maize area in North Bihar, in general and Zone I in particular was probably due to shift of area from kharif maize to paddy during post-green revolution period. In Zone II also, kharif maize area was shifted probably to rice but the increase in area under rabi maize has compensated more than the loss of area under kharif maize.

Barley occupied 3.15 per cent of gross cropped area in

Table 5.2

Zone-wise Share of Principal Crops in Cropping Pattern of North Bihar at the Three Points of Time (During Post-Green Revolution Period)

(in per cent)

Crop Period (Year Ending)	Rice	Wheat	Maize	Barley	Other Cereals	Arhar	Gram	Lentil	Other Pulses	Rape & Mustard	Other Oilseeds	Sugar-cane	Miscella-neous Crops
North Bihar													
1974	42.42	15.51	10.00	3.15	3.32	1.30	0.83	1.05	6.43	1.01	0.79	2.40	11.72
1984	43.53	18.92	10.04	0.98	1.41	0.82	0.54	0.96	7.22	0.98	0.92	2.06	11.55
1994	42.01	23.35	9.26	0.28	4.69	0.59	0.41	1.12	7.29	1.47	0.57	2.62	6.27
1974	42.40	17.47	11.77	3.84	2.17	1.79	0.96	1.18	5.3	0.73	0.86	3.45	7.80
1984	43.43	21.74	11.01	1.17	2.01	1.14	0.62	1.05	5.95	0.83	0.80	2.97	7.19
1994	43.26	26.22	9.69	0.36	4.13	0.82	0.45	1.24	6.07	1.30	0.72	3.82	1.85
1974	42.49	11.18	6.09	1.62	5.87	0.21	0.56	0.77	8.42	1.62	0.63	0.07	20.39
1984	43.74	12.96	7.98	0.58	0.13	0.13	0.39	0.78	9.88	1.30	1.17	0.13	20.75
1994	39.40	17.40	8.38	0.12	5.86	0.12	0.31	0.88	9.83	1.82	0.25	0.12	15.44

North Bihar in the year 1974 which declined continuously to 0.28 per cent in the year 1994. It was only due to shift of barley area to wheat area. In North Bihar, proportion of area under major pulses like *arhar* and gram to gross cropped area declined continuously during 1974-94. The share of lentil to gross cropped area also declined from 1.05 per cent in 1974 to 0.96 per cent in 1984 but increased to 1.12 per cent in 1994.

The proportion of area under other pulses to gross cropped area increased continuously from 6.43 per cent in 1974 to 7.29 per cent in 1994. Almost a similar trend was observed in both the zones under study. The increase in the proportion of area under other pulses was probably due to increase in area under summer pulses, particularly area under summer *moong* in the project area.

Rape and mustard is the major oilseed crop in the project area but its share in gross cropped area declined from 1.01 per-cent in 1974 to 0.98 per cent in 1984 but increased to 1.47 per cent in 1994 in North Bihar. The similar trend was observed in Zone II but in Zone I the share of rape and mustard increased continuously from 0.73 per cent in 1974 to 1.30 per cent in 1994. In North Bihar, the share of other oilseeds to gross cropped area increased from 0.79 per cent in 1974 to 0.92 per cent in 1984 but declined to 0.57 per cent in 1994. A similar trend was observed in Zone II, but the corresponding proportion declined continuously in Zone I, probably due to shift of area under other oilseeds to rape and mustard.

Sugarcane is the only cash crop which has widespread coverage across the districts of North Bihar. The share of sugarcane area to gross cropped area declined from 2.40 per cent in 1974 to 2.06 per cent in 1984 but increased to 2.62 per cent in 1994 in the project area. The similar trend was witnessed in Zone I but the share of sugarcane area to gross cropped area continuously increased in Zone II. Moreover, sugarcane has not been an important crop in Zone II. The decline in the importance of sugarcane in cropping pattern in the early period of green revolution was probably due to favourable economics of rice-wheat cropping sequences after the advent of HYVs of cereal crops.

An increase in area under rice, wheat, rape and mustard, and other oilseeds has been recorded whereas maize, pulses and sugarcane experienced a decline in area during the period 1993-94 to 1995-96 (Appendix II). Hence, it may be said that the cropping pattern did not respond to liberalisation in economy in the state of Bihar. It may be due to the fact that the impact of liberalisation is still to reach in remote areas in Bihar. However, the period of post-liberalisation is too short to draw conclusion on this particular aspect.

On the basis of above discussions, it may hardly be concluded that there has been shift in cropping pattern in the project area during the period under study. There was some differences between the cropping pattern of Zone I and Zone II also but it did not emerge clearly unless it was tested statistically.

Table 5.3

Zone-wise Estimated Coefficient of Concordance and Estimated X^2 Values of Cropping Pattern Followed at Specified Period of Time in North Bihar

Item of Concordance Test	*North Bihar*	*Zone I*	*Zone II*
Estimated Co-efficient of Concordance (W)	0.88	0.86	0.93
Estimated X^2 Values	21.12	20.64	22.32
Table Value (8 d.f.) and at 5 Per cent Level of Probability	15.5	15.5	15.5

The coefficient of Concordance Test was used to test the compatibility of cropping pattern followed at different points of time in North Bihar, Zone I and Zone II. An attempt was also made to test the compatibility of cropping pattern of Zone I and Zone II at three specified points of time. The value of calculated concordance coefficients (W), table value of X^2, 8, 0.05 and calculated X^2 value are presented in Table 5.3.

The calculated values of X^2 exceed the table value of X^2, 8 and 0.05. Hence the coefficients of concordance are significant at 5 per cent level of probability, indicating that the ranking of crops in the cropping patterns of North Bihar, Zone I and Zone II at three points of time (post-green revolution period) were incompatible. The ranks of crops in cropping pattern of Zone I and Zone II were also examined at three points of time (Table 5.4) which indicated a similar result, i.e., calculated values (X^2) were higher than the table value of X^2 (15.5). It clearly indicates that the cropping pattern of Zone I and Zone II were also incompatible at all the three points of time.

Table 5.4

Estimated Coefficient of Concordance and Estimated X^2 Values of Cropping Pattern Followed in Zone I and Zone II at Different Points of Time

Item of Concordance Test	*Year Ending 1974*	*Year Ending 1984*	*Year Ending 1994*
Estimated Coefficient of Concordance (W)	0.79	0.78	0.90
Estimated X^2 Values	18.96	18.72	21.6
Table Value (8 d.f.) and 5 per cent Level of Probability	15.5	15.5	15.5

The above findings clearly indicate that there has been a significant change in cropping pattern in North Bihar during post-green revolution period. It is true for Zone I and Zone II also. It may further be said that Zone I and Zone II also differ from each other as far as cropping pattern is concerned.

Changes in Area, Production and Productivity

To minimise the effect of fluctuations and get the normal data, the triennium average of area, production and productivity of principal crops grown in North Bihar were calculated at three points of time, i.e., ending 1974, 1984 and 1994. The period of study was divided in two time periods for comparative study. These periods are early period of green revolution (1974-84) and latter period of green revolution (1984-94). The relevant information is presented in Table 5.5.

It may be observed from the table that the area under total cereals and oilseeds recorded an increase in both the periods, i.e., 1974-84 and 1984-94, however, the decadal increase in area was comparatively higher under oilseeds (9.29% and 11.51%) than cereals (5.34% and 6.18%). The increase in area under cereals was mainly due to increase in area under rice and wheat during early period of green revolution (1974-84) whereas the corresponding increase during latter period of green revolution (1984-94) was only due to increase in area under wheat (2.40 lakh ha). Area under maize declined continuously during the period under study but the decadal decline was comparatively higher during 1974-84 (-39.26%) than the decline observed during 1984-94 (-5.56%). A decadal increase of 4.59 per cent in area under pulses was recorded during 1974-84 but a decadal decline of 3.90 per cent was observed during 1984-94. The slow growth in area under pulses in early period of green revolution and decline there of in the latter period of green revolution was probably due to decline in area under major pulse crops (*arhar* and gram) during the period under study. Oilseed crops recorded an increase in area throughout the period under study, however, the decadal increase was comparatively higher during

Table 5.5
Area, Production and Productivity of Principal Crops in North Bihar at Triennium Ending 1974, 1984 and 1994

Crop	*Year Ending 1974*	*Year Ending 1984*	*Year Ending 1994*	*Change from 1974-84 (%)*	*Change from 1984-94 (%)*
Area in Lakh ha					
Rice	19.25	20.69	20.50	7.48	-0.91
Wheat	7.04	9.00	11.39	27.71	26.61
Maize	7.87	4.78	4.51	-39.26	-5.56
Total Cereals	33.79	35.60	37.80	5.34	6.18
Total Pulses	4.39	4.59	4.41	4.59	-3.90
Total Oilseeds	0.83	0.91	1.01	9.24	11.51
Sugarcane	1.10	0.99	1.28	-10.66	29.75
Production in Lakh Tonnes					
Rice	13.11	17.20	18.98	31.16	10.35
Wheat	8.57	12.95	21.41	51.09	65.31
Maize	4.64	5.68	9.03	22.30	59.08
Total Cereals	30.29	35.99	51.31	18.79	42.55
Total Pulses	1.60	2.03	2.15	26.66	6.22
Total Oilseeds	0.37	0.57	0.76	53.84	31.65
Sugarcane	40.13	33.03	60.61	-17.70	83.48
Yield in kg/ha					
Rice	681	831	926	22.02	11.43
Wheat	1216	1439	1878	18.33	30.50
Maize	590	1188	2002	101.35	68.51
Total Cereals	896	1010	1357	12.72	34.35
Total Pulses	364	441	488	21.15	10.65
Total Oilseeds	450	634	748	40.88	17.98
Sugarcane	36195	33514	47154	-7.40	40.69

1984-94 (11.51%) than the period 1974-84 (9.24%). Area under sugarcane had a setback during 1974-84 when a decadal decline of 10.66 per cent was observed but the area under sugarcane increased to all time height of 1.28 lakh ha in North Bihar in the year 1994 and recorded a decadal increase of nearly 29.75 per cent during the period 1984-94.

Zone-wise analysis revealed the incongruence between the zones and vis-a-vis North Bihar with respect to variation in areas of different crops under study (Table 5.6). Area under rice increased in Zone I during both the periods though the increase was moderate, i.e., 5.65 per cent in the early period of green revolution and 1.74 per cent in latter period of green revolution. On the other hand, area under rice increased by 11.68 per cent in Zone II during the early period but declined by -6.57 per cent during the latter period. It was probably due to a sharp decline in area under summer/winter rice and upland rice in Zone II. Sinha (1991) has also observed the decline of 2.37 per cent in summer rice acreage in North Bihar during 1980-81 to 1990-91.

It is worth pointing out that the cultivation of summer rice was adopted in a big way in Zone II during the period preceding to the green revolution which could not sustain in the eightieth decade probably due to its higher requirement for irrigation water which was adversely affected by inadequate and erratic electric supply, on the one hand and inadequate availability of fuel (diesel) in the area, on the other.

Wheat is the only crop which recorded a sustained area growth of more than 25 per cent in the project area during the period under study. The similar trend was observed in both zones but Zone II witnessed comparatively higher decadal growth in area under wheat (32.27%) during the latter period of green revolution (1984-94). In North Bihar, area under maize declined in both the periods but the decline was comparatively higher in early period of green revolution (-39.26%) than that of

Table 5.6

Area, Production and Productivity of Principal Crops in Zone I and Zone II at Triennium Ending 19974, 1984 and 1994.

Crop	Zone I					Zone II				
	Ending	Ending	Ending	Decadal Change	Decadal Change	Ending	Ending	Ending	Decadal	Decadal
	1974	1984	1994	1984 over 1974(%)	1994 over 1984(%)	1974	1984	1994	1984 over 1974(%)	1994 over 1984(%)
Area in Lakh ha										
Rice	13.25	14.00	14.24	5.65	1.74	5.99	6.69	6.25	11.68	-6.57
Wheat	5.46	7.01	8.63	28.51	23.03	1.58	1.98	2.76	24.94	39.27
Maize	7.91	3.55	3.19	-55.08	-10.10	0.86	1.22	1.33	42.24	9.10
Cereals	24.27	25.61	27.53	5.51	7.47	9.51	9.98	10.27	4.9	2.89
Pulses	2.96	2.86	2.84	-3.65	-0.41	1.42	1.73	1.56	21.76	-9.80
Oilseeds	0.50	0.53	0.67	5.51	27.60	0.33	0.38	0.33	14.92	-10.90
Sugarcane	1.08	0.96	1.26	-1.15	31.16	0.019	0.026	0.021	35.59	-20.57
Production in Lakh Tonnes										
Rice	9.58	12.02	13.92	25.41	15.78	3.52	5.18	5.06	46.76	-2.24
Wheat	6.57	10.31	16.08	56.94	55.99	2.01	2.64	5.32	30.92	101.68

Contd...

Contd...

Maize	3.92	4.44	6.60	13.28	48.65	0.49	1.23	2.42	149.14	96.63
Cereals	22.79	27.56	38.90	20.91	41.15	7.50	8.42	12.40	12.35	47.15
Pulses	0.90	1.20	1.29	33.71	6.66	0.69	0.82	0.86	17.54	5.56
Oilseeds	0.22	0.37	0.53	64.74	42.55	0.14	0.19	0.22	35.06	12.50
Sugarcane	39.46	32.31	59.47	-18.10	84.04	0.67	0.71	1.13	5.70	58.33
Yield in kg/ha										
Rice	723	858	977	18.67	13.86	588	774	809	31.63	4.52
Wheat	1203	1469	1863	22.11	26.82	1270	1331	1928	4.80	44.85
Maize	495	1249	20.66	152.32	65.41	576	1010	1820	74.34	80.19
Cereals	938	1075	1413	14.60	31.44	788	844	1207	7.10	43.00
Pulses	304	422	452	38.81	7.10	490	473	554	-3.46	17.12
Oilseeds	455	710	794	56.04	11.83	442	520	657	17.64	26.34
Sugarcane	36234	33535	47056	-7.44	40.31	34060	26551	52933	22.04	99.37

latter period of green revolution (-56%). A similar declining trend in area under maize was observed in Zone I but the trend was just reverse in Zone II where area under maize increased during both the periods under study. The increase was comparatively higher during early period (42.24%) than the latter period of green revolution (9.10%). It is worth mentioning that the increase in area under maize was mainly due to adoption of winter maize in the area (Appendix III).

Area under pulses also increased by 4.59 per cent during the period 1974-84 but declined by -3.90 per cent during the period 1984-94. The similar trend was observed in Zone II, however, the increase during the early period and decline during latter period in pulses area was comparatively higher in Zone II than that of corresponding increase/decrease observed in North Bihar. In Zone I, area under pulses declined during both the periods but the decline in area under pulses was moderate, that is -3.65 per cent during the early period and only -0.41 per cent in the latter period.

During the early period (1974-84), both the zones under study witnessed increase in oilseeds area but the increase was comparatively higher in Zone II (14.92%) than Zone I (5.51%). On the other hand, Zone II witnessed a decline of -10.90 per cent in area under oilseeds, during the latter period (1984-94) whereas there was sharp increase under oilseeds area in Zone I (27.60%) during the period 1984-94.

In North Bihar, area under sugarcane declined by -10.66 per cent during 1974-84, but recorded a decadal increase of 29.75 per cent during 1984-94. Zone I, being major region of sugarcane cultivation, witnessed a marginal decline of -1.15 per cent in area under sugarcane during 1974-84 but witnessed an increase of 31.16 per cent during 1984-94, whereas just reverse trend was observed in Zone II, i.e., increase during 1974-84 (35.59%) and decline of -20.57 per cent during 1984-94.

Production

In North Bihar, production of all principal crops, viz., rice, wheat, maize, pulses and oilseeds increased during the period under study but rice, pulses and oilseeds recorded comparatively higher increase during the early period of green revolution (1974-84) than the latter period of green revolution (1984-94) whereas wheat and maize recorded comparatively higher increase in their productions during the latter period (1984-94) than that of early period (1974-84). It is worth pointing out that the production of wheat exceeded the production of rice in North Bihar at the triennium ending 1994. Production of sugarcane declined during 1974-84 (-19.70%), but a sharp increase was observed during the latter period mainly caused by an increase in area and productivity of sugarcane during the period.

In Zone I, production of all the principal crops namely, rice, wheat, maize, pulses and oilseeds recorded increase during both the periods under study, however, the increase was more pronounced in case of wheat (56.94%) and oilseeds (64.74%) during 1974-84. Production of these two crops maintained the increasing trend during 1984-94 also. Maize production which had growth of only 13.28 per cent during the early period of green revolution recorded decadal increase of 48.65 per cent during 1984-94. The growth in maize production during the period understudy was mainly due to increase in the productivity of maize which was made possible due to introduction of winter maize in the project area. In Zone I, the pulses and oilseed production increased by 6.66 per cent and 42.55 per cent respectively during 1984-94 which were less than the corresponding increase achieved during the period 1974-84. Production of sugarcane declined by -18.10 per cent, during 1974-84 which increased by 84.04 per cent during 1984-94.

In Zone II, rice production increased by 46.76 per cent during the period 1974-84, mainly due to increase in productivity of the crop (31.63%). After recording a sharp growth in rice production during 1974-84, Zone II witnessed a decline of -2.24 per cent in rice production due to decline in area (-6.57) during the period. Production of all principal crops under study (except rice) recorded growth in production during both the periods. Production of wheat and sugarcane witnessed comparatively higher rate of growth during the period 1984-94 than the corresponding growth recorded during 1974-84, whereas the production of rice, maize, pulses and oilseeds recorded comparatively higher growth during 1974-84 than that of growth observed during the period 1984-94.

Productivity

In North Bihar, productivity of all principal crops increased during the period under study. Moreover, the productivity of sugarcane declined during the period 1974-84. Contrary to general belief of stagnant growth in agricultural production and productivity, rice and wheat maintained the increasing trend in production and productivity in North Bihar during the period understudy (1974-94). The productivity of these crops was comparatively higher, during the period 1984-94 than that of productivity of these crops observed during the period 1974-84, indicating an increasing trend in productivity of principal cereal crops. In addition to this, productivities of sugarcane also grew by 40.69 per cent during 1984-94, whereas the productivity of this crop declined by -7.40 per cent in the preceding period (1974-84). Productivity of maize, pulses and oilseeds grew at faster rate during the period 1974-84 than the corresponding growth rate recorded during 1984-94. The decadal increase in productivity of rice, maize, pulses and oilseeds was comparatively higher during 1974-84 than corresponding increase recorded during 1984-94 in Zone I. The situation was just reversed in case of Zone II, i.e., the comparatively higher decadal increase in productivity of these crops (except rice) during 1984-94 than that of corresponding increase observed during 1974-84. Moreover, in North Bihar including Zone I and Zone II, rice witnessed comparatively lower decadal increase in productivity during 1984-94 than that of 1974-84. In both the zones, decadal increase in wheat productivities was comparatively higher during 1984-94 than the corresponding increase observed during 1974-84. It is worth pointing out that the increase in wheat productivity was much lower (4.80%) in Zone II during the period 1974-84. In Zone I, maize productivity recorded comparatively higher growth during both the periods, however, it was much higher during 1974-84 (152.32%).

Sugarcane productivity was comparatively higher during the period 1984-94, however, the growth in sugarcane productivity was much higher in Zone II (99.37%) than the corresponding growth recorded in Zone I (40.31%). As has earlier been explained, sugarcane productivity declined during the period 1974-84 in North Bihar and a similar trend was observed in both the zones, however, the decline was comparatively higher in Zone II (-22.04%) than the decline (-7.44%) observed in Zone I.

On the basis of above discussions, it may be concluded that there has been a substantial increase in the production of principal crops in North Bihar, however, the increase in production of wheat and oilseeds was more pronounced in both the periods. There was spectacular growth in maize production during the period 1984-94 mainly due to introduction of winter maize in the project area. Rice production also increased during the period 1974-84 but the increase experienced a setback during the period 1984-94. Like rice production, pulses also witnessed the same fate. There was also a spectacular increase in sugarcane production during 1984-94 which had experienced a considerable decline during the early period of green revolution.

An increase in production of wheat, oilseeds and sugarcane was due to increase in area and productivity whereas the increase in production of other principal crop categories was only due to increase in productivity. It is also worth pointing out that the productivities of almost all principal crop categories were comparatively higher during 1984-94 than 1974-84 in Zone II, where the situation was just reverse in Zone I. Moreover wheat showed increasing trend in productivity in both the zones and rice observed a decline in productivity, during 1984-94 over period 1974-84.

Growth Performance

Major Crop Categories

In this section, an attempt has been made to analyse the compound annual growth rates in area, production and productivity of principal crop categories/crops, during specified periods, i.e., study period (1970-71 to 1993-94), Period I (1970-71 to 1979-80), Period II (1980-81 to 1989-90) and Period III (1990-91 to 1993-94). Zone-wise annual growth rates of area, productivity and production of major crop categories in North Bihar (Zone I and Zone II) during specified periods are presented in Tables 5.7, 5.8 and 5.9 respectively.

It may be observed from the Table 5.7 that the annual growth rate of area under all the major crop categories increased during the study period, however, the foodgrains recorded comparatively higher increase in area growth rate (0.6%) than the corresponding increase in cereals and oilseeds (0.4% each) and pulses (0.1%). Annual growth rate of area under foodgrains was comparatively higher in Zone I (0.7%)j than the corresponding growth achieved in Zone II (0.5%) during the period under study. Area growth rate of cereals was 0.6 per cent in Zone I and 0.4 per cent in Zone II during the study period but there was a decline in area under pulses in Zone I (-0.1%). The Zone II, no doubt, achieved the annual growth of 0.5 per cent in pulses area during the period.

Table 5.7

Zone-wise Annual Growth Rates in Area of Major Crop Categories of North Bihar During Study Period and Sub-Periods

(in per cent)

Crop Categories	*Study Period (1970-71 to 1993-94)*	*Sub-Periods*		
		Period-I (1970-71 to 1979-80)	*Period-II (1980-81 to 1989-90)*	*Period-III (1990-91 to 1993-94)*
North Bihar				
Total Cereals	0.4	1.3	0.11	2.1
Total Pulses	0.1	1.2	0.2	7.4
Total Foodgrains	0.6	2.0	2.9	2.6
Total Oilseeds	0.4	2.6	-0.3	-2.9
Zone I				
Total Cereals	0.5	1.8	0.5	0.7
Total Pulses	-0.1	0.3	0.2	2.1
Total Foodgrains	0.7	2.1	4.1	0.1
Total Oilseeds	0.7	2.4	0.2	-4.5
Zone II				
Total Cereals	0.4	-0.3	1.3	7.6
Total Pulses	0.5	3.1	0.1	21.1
Total Foodgrains	0.5	1.7	1.2	6.9
Total Oilseeds	-0.1	3.0	-1.1	0.1

During specified sub-periods of green revolution, annual growth rate in area under foodgrains increased from 2 per cent during Period I to 2.9 per cent during Period II but declined to 2.6 per cent during Period III in North Bihar. Moreover, there was absolute increase in area under foodgrains during Period III

(2.02 lakh ha) which was comparatively higher than the increase of 2.00 lakh ha during the Period II.

The lower growth observed in area under foodgrain during Period III than the growth rate observed during Period II was due to comparatively higher base of Period III than that of Period II. Annual growth rate in area under cereals and pulses declined from 1.3 per cent and 1.2 per cent during Period I to 0.1 per cent and 0.2 per cent during Period II but increased to 2.1 per cent and 7.4 per cent during Period III respectively.

Annual growth rate in area under foodgrains in Zone I followed almost similar trend, i.e., increase in growth rate during Period II and a decline in growth rate during Period III. On the other hand, the reverse trend was observed in Zone II, i.e., a decline in growth rate during Period II and increase during Period III. Despite the decline in net area sown, the growth in area under foodgrains showed an increasing trend throughout the period under study and an accelerated growth in Zone II during Period III. Contrary to the general trend of declining area under foodgrains in the country, the area under foodgrains recorded an increasing trend in North Bihar, which has been probably to meet the food requirements in foodgrain deficit regions like North Bihar.

In North Bihar, annual growth rate in area under oilseeds was 0.4 per cent during the study period, however, Zone I recorded comparatively higher growth rate in oilseeds area (0.7%) during the period under study, but Zone II witnessed a decline in area under oilseeds (-0.1%) during the period. During Period I, growth rate in area under oilseeds in North Bihar, Zone I and Zone II was positive and comparatively higher than the respective growth rate obtained during Period II and Period III. It is worth pointing that North Bihar and Zone I witnessed the decline in area under oilseeds during Period II and Period III but there was marginal increase in area under oilseeds in Zone II during Period III (0.1%).

Introduction of bio-chemical technology in Indian agriculture has influenced the productivity of principal crops but annual growth rate in productivity of most of the crops could not be maintained, particularly after mid-1980s. The present analysis is planned to test whether the growth rate in productivity of principal crops attained in early period of green revolution could sustain during latter part of green revolution in North Bihar.

In North Bihar, annual growth rate in productivity of foodgrains was 2.1 per cent during the study period, however, the productivity growth rate was estimated to 1.2 per cent during Period I which increased to 1.6 per cent during Period II and maintained at the same level during Period III. During the period under study, Zone II achieved comparatively higher annual growth rate (2.3%) than that of Zone I (2.1%). In Zone I, there was declining growth rate in productivity of foodgrains from 0.7 per cent during Period I to 0.4 per cent during Period II and the productivity growth rate turned to be negative (-0.5%) during Period III. On the other hand, annual growth rate in foodgrain productivity showed increasing trend in Zone II from 2.6 per cent during Period I to 4 per cent during Period II and 9.9 per cent during Period III.

Annual growth rates in cereal productivity were 2.2 per cent in North Bihar, 2.1 per cent in Zone I and 1.9 per cent in Zone II. An analysis of cereal productivity in different sub-periods indicated an increase in productivity growth rate from Period I to Period II which could not be sustained during Period III in North Bihar and Zone I. On the other hand, in Zone II, there was a decline in cereal productivity growth rate from Period I (4.1%) to Period II (3.9%) but increased to 9.9 per cent in Period III. Despite the comparatively higher growth rate in

Table 5.8

Zone-wise Annual Growth Rate in Productivity of Major Crop Categories of North Bihar During Study Period and Sub-Periods

(in per cent)

Crop Categories	*Study Period (1970-71 to 1993-94)*	*Sub-Periods*		
		Period-I (1970-71 to 1979-80)	*Period-II (1980-81 to 1989-90)*	*Period-III (1990-91 to 1993-94)*
North Bihar				
Total Cereals	2.2	2.1*	3.5	2.6
Total Pulses	0.08	0.4	-1.0	0.2
Total Foodgrains	2.1	1.2	1.6	1.6
Total Oilseeds	2.7	0.1	1.9	5.9
Zone I				
Total Cereals	2.1	1.4	3.3	1.9
Total Pulses	3.4**	10.9	-1.6	-0.2
Total Foodgrains	2.1	0.7	0.4	-0.5
Total Oilseeds	3.1	0.5	2.1**	9.6
Zone II				
Total Cereals	1.9	4.1	3.9	9.9
Total Pulses	0.4***	-0.7	-0.3	3.2
Total Foodgrains	2.3	2.6	4.0	9.9
Total Oilseeds	1.9**	-0.08	0.08**	0.2**

* Significant at 10 per cent level of probability.
** Significant at 5 per cent level of probability.
*** Significant at 1 per cent level of probability.

cereal productivity in Zone II, it was not found statistically significant. But the cereal productivity increased significantly in North Bihar during Period I. During the study period, annual growth rate in pulse productivity was 0.1 per cent in North Bihar. The growth rates of pulse productivity was comparatively higher in Zone I (3.4%) than Zone II (0.4%) and these were statistically significant. During sub-periods, the pulse productivity growth rate was negative during Period II in North Bihar and Zone I and during Period I and Period II in Zone II, indicating fluctuation of pulse production in the project area.

Oilseeds crops are, no doubt, minor crops in North Bihar but this category of crops achieved comparatively higher growth rate in productivity (2.7%) than the productivity growth rates of other categories of crops during the study period. Zone I achieved comparatively higher growth rate in oilseed productivity (3.1%) than the corresponding rate of growth in Zone II (1.9%) during study period but the latter was found statistically significant at 5 per cent level of probability. Sub-period-wise analysis of oilseed productivity indicated an increasing trend in North Bihar and both the zones under study. Moreover, Zone II witnessed negative growth rate in oilseed productivity during Period I but achieved statistically significant growth rate during Period II and Period III.

Annual growth rate in the production of major crops categories have also been studied for the study period as well as sub-periods. In North Bihar annual growth rate in production of foodgrain was estimated to 2.7 per cent during study period (Fig. I) but it was 3.3 per cent in Period I which increased to 4.5 per cent in Period II but declined to 4.2 per cent in Period III. The increase in growth rate of foodgrain production in Period II over Period I was due to increase in annual growth rate in area and productivity of foodgrain during Period II, but a decline in production growth rate during Period III was due to decline in area growth rate during the period. It is worth pointing out that there was a significant increase in foodgrain production (2.7%) in Zone I during the study period but it increased from 2.9 per cent in Period I to 4.5 per cent in Period II and remained almost stagnant (zero growth) in Period III. It so happened due to decline in annual growth rate in productivity of foodgrains in Zone I during the subsequent sub-periods. In Zone II, there was

Table 5.9
Zone-wise Annual Growth Rate in Production of Major Crop Categories of North Bihar During Study Period and Sub-Periods

(in per cent)

Crop Categories	*Study Period (1970-71 to 1993-94)*	*Sub-Periods*		
		Period-I (1970-71 to 1979-80)	*Period-II (1980-81 to 1989-90)*	*Period-III (1990-91 to 1993-94)*
North Bihar				
Total Cereals	2.6	3.2*	4.1	5.7
Total Pulses	1.1	1.8	0.08	7.2
Total Foodgrains	2.7	3.3	4.5	4.2
Total Oilseeds	3.2	2.4	1.6	3.0
Zone I				
Total Cereals	2.6	3.0	4.0	2.8
Total Pulses	3.4**	11.5	-1.6	2.6
Total Foodgrains	2.7***	2.9	4.5	0.0
Total Oilseeds	3.9	2.7	2.4**	4.5
Zone *II*				
Total Cereals	2.3**	3.9	5.2	18.3
Total Pulses	0.08***	2.3	0.0	17.2***
Total Foodgrains	2.8	4.5	5.1	18.0
Total Oilseeds	1.7	2.3	-0.2**	0.5**

* Significant at 10 per cent level of probability.
** Significant at 5 per cent level of probability.
*** Significant at 1 per cent level of probability.

a comparatively higher growth rate in foodgrain production during study period (2.8%), but it was not statistically significant. Moreover, the growth rate in foodgrain production showed increasing trend from 4.5 per cent in Period I to 5.1 per cent in Period II and 18 per cent in Period III. It was due to increasing growth rate in productivity and substantial growth in area of foodgrains in the zone. Amongst the foodgrains, cereals which constituted nearly 96 per cent of foodgrain production in North Bihar grew at a rate of 2.6 per cent during the study period (Figure I) and its growth rate showed increasing trend from 3.2 per cent in Period I to 4.1 per cent in Period II and 5.7 per cent in Period III. Moreover, the annual growth rate in cereal production was comparatively lower in Period I than that of subsequent periods, but it was statistically significant at 10 per cent level of probability. It is worth pointing out that the cereal production growth rate was positive during all the three sub-periods but the increasing trend in its growth rate could not be maintained in Zone I, whereas Zone II witnessed significant growth rate in cereal production during study period and it maintained increasing trend during sub-periods, particularly due to increasing productivity growth rates of cereals during subsequent periods.

In North Bihar, annual growth rate in pulse production was estimated to 1.1 per cent during the study period (Figure I), however, it was 3.4 per cent in Zone I and 0.1 percent in Zone II. The growth rate in pulse production achieved during study period was statistically significant in Zone I and Zone II. In North Bihar, growth rate in pulse production declined from 1.8 per cent during Period I to 0.1 per cent during Period II, but increased to 7.2 per cent during Period III. In Zone I, production growth rate not only declined but turned to be negative from Period I (11.5%) to Period II (-1.6%) and again increased to 2.6 per cent during Period III. In Zone II also, production growth rate declined from Period I to Period II and increased to much higher level of 17.2 per cent which was also found significant at 1 per cent level of probability. Moreover, the increase in pulse production during Period III was due to increase in area under pulses (21.1%) and their productivity growth rate (3.2%).

Annual growth rate in oilseed production was also positive

FIGURE 1

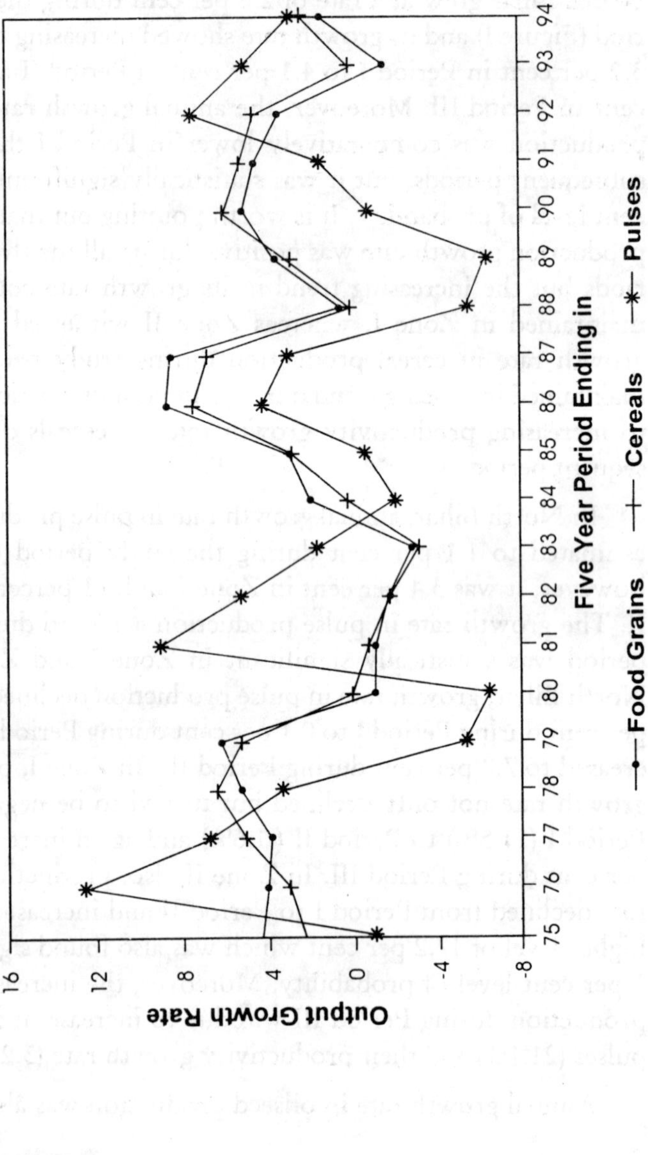
Output Growth Rate
16
12
8
4
0
-4
-8
75 76 77 78 79 80 81 82 83 84 85 86 87 88 89 90 91 92 93 94
Five Year Period Ending in
Food Grains
Cereals
Pulses

FIGURE 2

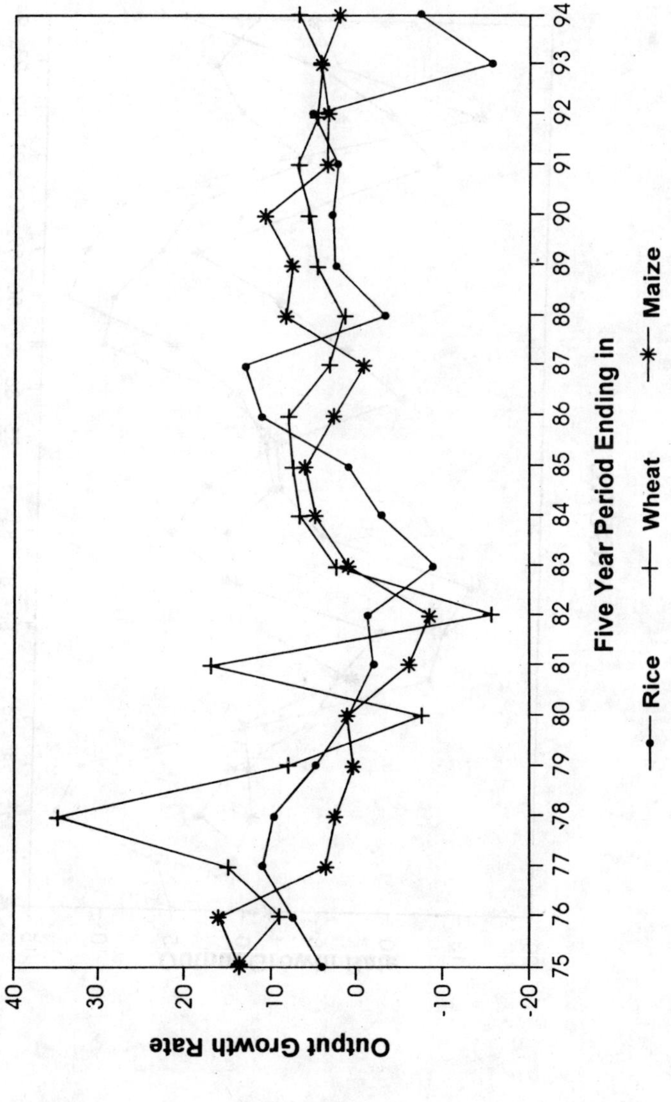
Output Growth Rate
40
30
20
10
0
-10
-20
75
76
77
78
79
80
81
82
83
84
85
86
87
88
89
90
91
92
93
94
Five Year Period Ending in
Rice
Wheat
Maize

FIGURE 3

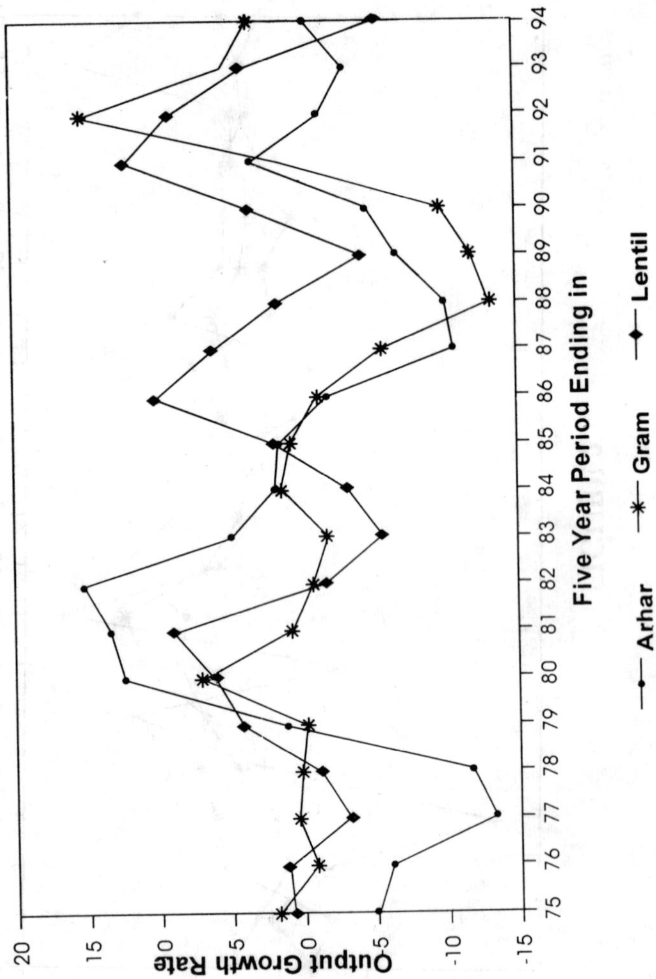
Output Growth Rate
20
15
10
5
0
-5
-10
-15
75
76
77
78
79
80
81
82
83
84
85
86
87
88
89
90
91
92
93
94
Five Year Period Ending in
Arhar
Gram
Lentil

FIGURE 4

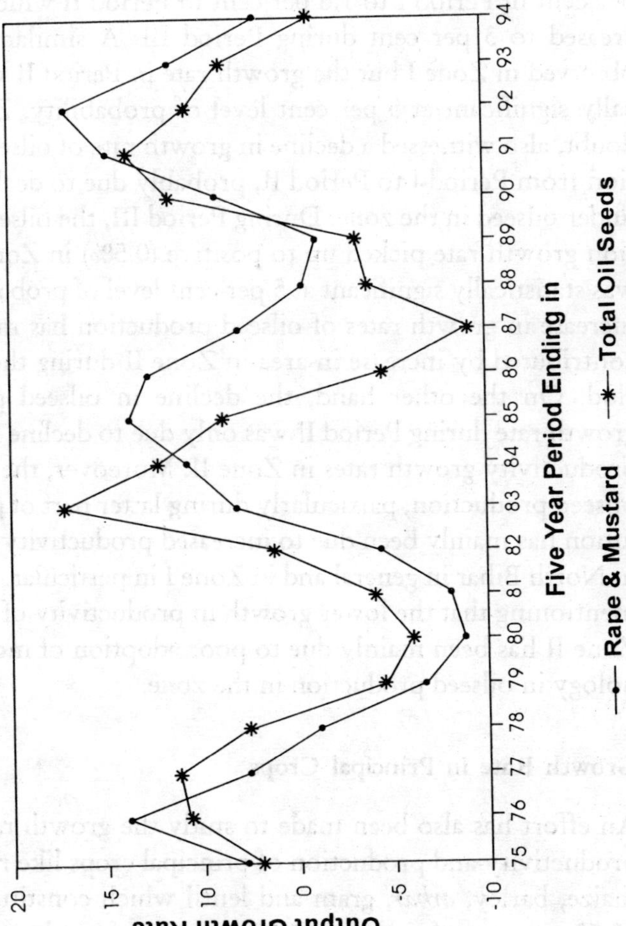
Output Growth Rate
20
15
10
5
0
-5
-10
75 76 77 78 79 80 81 82 83 84 85 86 87 88 89 90 91 92 93 94
Five Year Period Ending in
Rape & Mustard
Total Oil Seeds

(3.2%) in North Bihar (Figure IV) during the study period (1970-71 to 1993-94). During sub-periods, it declined from 2.4 per cent in Period I to 1.6 per cent in Period II which again increased to 3 per cent during Period III. A similar trend was observed in Zone I but the growth rate in Period II was statistically significant at 5 per cent level of probability. Zone II, no doubt, also witnessed a decline in growth rate of oilseed production from Period I to Period II, probably due to decline in area under oilseed in the zone. During Period III, the oilseed production growth rate picked up to positive (0.5%) in Zone II and it was statistically significant at 5 per cent level of probability. The increase in growth rates of oilseed production has mainly been contributed by increase in area in Zone II during the study period. On the other hand, the decline in oilseed production growth rate during Period II was only due to decline in area and productivity growth rates in Zone II. Moreover, the growth in oilseed production, particularly during latter part of green revolution has mainly been due to increased productivity of oilseeds in North Bihar in general and in Zone I in particular. It is worth mentioning that the lower growth in productivity of oilseeds in Zone II has been mainly due to poor adoption of modern technology in oilseed production in the zone.

Growth Rate in Principal Crops

An effort has also been made to study the growth rate in area, productivity and production of principal crops like rice, wheat, maize, barley, *arhar*, gram and lentil which constituted nearly 90.53 per cent of area and 95.39 per cent of production of total foodgrains in the project area (Tables 5.10, 5.11 and 5.12 respectively).

Rice is still a major crop with respect to area which occupies nearly 49.36 per cent of foodgrain cropped area in North Bihar. Annual growth rate in area under rice was 0.3 per cent during the study period, however, it was comparatively higher (1.5%) during Period I. The area under rice remained almost stagnant during Period II, whereas a marginal increase in area under rice (0.2%) was observed during Period III. Zone I had the identical annual growth rate in area under rice during study period and maintained increasing growth rate during 1970s and 1980s (Period I and Period II) but witnessed a declining growth rate (-1.8%) during the early 1990s (1990-91 to 1993-94). Zone I recorded a comparative higher growth rate in area under rice during the study period (0.4%) and during Period I (3.1%) than the corresponding growth observed in North Bihar and Zone I but recorded a negative growth rate in Period II (-0.3%) and much higher growth rate (5.2%) during Period III. An increase in area under rice, during early 1990s has probably been due to adoption of boro-rice cultivation in the area.

Wheat is now the most important foodgrain crop in North Bihar since its production exceeded the production of traditional staple food crop of rice. Annual growth rate in area of wheat was 2.2 per cent during study period in North Bihar, however, the area growth rate was much higher during Period I (5.7%) which declined to 3.2 per cent during Period II but again increased to 4.5 per cent during Period III. In Zone I, annual growth rate in wheat area was 2.3 per cent during the study period and it showed a declining trend from 6.9 per cent in Period I to 2.9 per cent in Period II and 1.6 per cent in Period III. Moreover, increase in area growth rate during Period III was statistically significant at 10 per cent level of probability. Zone II witnessed a growth rate of 2.4 per cent in area under wheat during study period but an increasing trend was observed over the periods under study and it increased to 14 per cent in Period III which was also statistically significant at 10 per cent level of probability.

Table 5.10

Zone-wise Annual Growth Rate in Area of Principal Crops of North Bihar During Study Period and Sub-Periods

(in per cent)

Crop Categories	*Study Period (1970-71 to 1993-94)*	*Sub-Periods*		
		Period-I (1970-71 to 1979-80)	*Period-II (1980-81 to 1989-90)*	*Period-III (1990-91 to 1993-94)*
North Bihar				
Rice	0.3	1.5	0.00	0.2
Wheat	2.2	5.7	3.2	4.5
Maize	-0.7*	3.5	2.2**	1.9
Barley	-11.10	-10.8	-13.3*	-0.08
Arhar	-3.0	-4.4	-4.5	4.7
Gram	-3.3***	-1.2	-4.2	-2.6
Lentil	-0.2	0.4	0.7	0.08
Sugarcane	0.1	-1.3	-1.2	-1.4
Rape and Mustard	0.4	2.6	-0.3	-2.9
Zone I				
Rice	0.3	0.08	0.4	-1.8
Wheat	2.3	6.9	2.9	1.6*
Maize	-2.9	-6.5	-3.1**	-0.2
Barley	-11.1	-11.2	13.1***	-1.4
Arhar	-3.2	-4.5	-4.1	4.5
Gram	-3.6	-2.6	-3.4	-2.5
Lentil	-0.7	0.0	0.2	-0.2
Sugarcane	0.2	-1.5	-1.3	-1.4
Rape and Mustard	0.7	2.4	0.2	-4.5
Zone II				
Rice	0.4	3.1	-0.3	5.2

Contd...

Contd...

Wheat	2.4	2.9	4.3	14.0*
Maize	1.5	5.7	-0.5	8.9
Barley	-11.2*	-8.3***	-14.9	0.08
Arhar	-2.1	-1.0*	-0.7	43.5
Gram	-2.3	2.7	-6.3**	23.6
Lentil	1.6	2.1	2.1	5.2
Sugarcane	-2.0	4.3	-14.7	-2.5
Rape and Mustard	-0.1	3.0	-1.1	0.80

* Significant at 10 per cent level of probability.
** Significant at 5 per cent level of probability.
*** Significant at 1 per cent level of probability.

In North Bihar, area growth rate of maize showed negative trend (-0.7%) during the study period and it was statistically significant at 10 per cent level of probability. During sub-periods under study, the growth rate in area under maize was positive but showed declining trend from 3.5 per cent in Period I to 2.2 per cent in Period II and 1.9 per cent in Period III. Moreover, the area growth rate obtained in Period II was statistically significant at 5 per cent level of probability. Zone I witnessed negative growth in area under maize during the study period as well as in all the subsequent sub-periods but the decline was statistically significant only during Period II. On the other hand, Zone II recorded a positive growth rate of 1.5 per cent in area under maize during the study period. During Period I, growth rate in area under maize was 5.7 per cent which declined to -0.5 per cent during Period II but again increased to 8.9 per cent in Period III.

The negative growth rate in area under maize during the study period in North Bihar was only due to decline in area under kharif maize which could not be compensated by increasing area under rabi maize. On the other hand, Zone II maintained the positive growth rate in area under maize (except during Period II), probably due to increase in area under rabi maize which has been higher than the decline in area under kharif maize in the region (Appendix III).

Barley has been the most sufferer in the post-green revolution period, particularly with respect to coverage. Area growth rate of barley showed negative trend during the study period in North Bihar. A similar trend was observed in both the zones under study. However, the decline was statistically significant during Period II in North Bihar and Zone I, and study period and Period I in Zone II. This was mainly due to shift of barley area to area under wheat cultivation.

Area growth rate of *arhar* was negative (-3%) during the study period in North Bihar which remained negative during Period I (-4.4%) and Period II (-4.5%) but turned to be positive during Period III. Zone I and Zone II followed the similar trend with respect to annual growth in area under *arhar* but Zone II achieved a much higher annual growth rate of 43.5 per cent during Period III but it was not statistically significant.

In North Bihar, area under gram witnessed a negative growth rate (-3.3%) during the study period which was statistically significant at 1 per cent level of probability. The growth rate in area under gram remained negative in all the three subsequent periods under study but the rate of decline slowed down during Period III as compared to Period II. The similar trend was observed in Zone I also. A negative growth rate in area under gram was observed in Zone II during the study period but a positive growth of 2.7 per cent was recorded during Period I and much higher growth of 23.6 per cent during Period III. Area under gram declined by -6.3 per cent during Period II in Zone II and it was statistically significant at 5 per cent level of probability.

Annual growth rate in area under lentil was negative (-0.2%)

during the study period in North Bihar but it showed a positive growth rate in subsequent sub-periods, probably due to substantial growth rate in area under lentil in Zone II throughout the period under study. On the other hand, Zone I observed a negative growth rate during the study period (-0.7%). Area under lentil did not show any growth during Period I, but a positive growth rate of 0.2 per cent during Period II which declined with the same rate during Period III.

On the whole, North Bihar experienced a decline in area under major pulses during post-green revolution period which was mainly contributed by comparatively higher decline in area in Zone I. Zone II performed comparatively better through bringing more area under pulses particularly during early period of present decade (Period III).

In North Bihar, area under rape and mustard recorded annual compound growth rate of 0.4 per cent during the study period, however, comparatively higher growth in area was observed in Zone I (0.7%) but Zone II witnessed a negative growth rate of -0.1 per cent during the period. During sub-periods under study there was a declining trend in area growth rates of rape and mustard which turned to be negative in Period II (-0.3%) which further declined to -2.9 per cent during Period III. similar declining trend was observed in Zone I during sub-periods, but, the negative growth rate was observed only during Period III. Moreover, in Zone II, the growth rate declined from Period I (3.0%) to Period II (-1.1%) but further increased to 0.80 per cent during Period III.

During the study period, area under sugarcane increased by 0.1 per cent and 0.2 per cent in North Bihar and Zone I respectively, but a decline of -2.0 per cent was observed in Zone II. During all the three sub-periods, North Bihar and Zone I witnessed a decline in area under sugarcane but an increase of 4.3 per cent in area under sugarcane was observed during Period I in

Zone II. Moreover, a much higher rate of decline (-14.7%) was observed in this particular zone during Period II and a moderate decline of -2.5 per cent during Period III.

Growth in Productivity

It was expected to have increasing growth rate in productivity of principal crops during post-green revolution period. Keeping in view the above hypothesis, an effort has been made in present sub-section to find out trend in productivity growth rate of principal crops in the project area (Table 5.11). Despite the increase in per ha rice productivity from 786 kg during Period I to 917 kg during Period III in North Bihar, the annual growth in rice productivity showed a declining trend (-0.2%) during the study period but it was not statistically significant. It was probably due to recurrent fluctuations in area and production of rice which has been mainly contributed by the natural hazards like floods and droughts in the area. Sub-period-wise analysis, however, indicated a growth of 0.4 per cent in rice productivity during Period I but it was not found statistically significant. During Period II and Period III, annual growth rate in rice productivity registered much declining trend, being -8.2 per cent and -4.3 per cent respectively. On the other hand, Zone I and Zone II witnessed increasing trend in rice productivity and registered growth rate of 1.5 per cent and 1.2 per cent respectively during the study period. During Period I and Period II annual growth in rice productivity showed an increasing trend of 3.3 per cent and 3.5 per cent in Zone I and 3.5 per cent and 4.6 per cent in Zone II respectively. However, both the zones registered declining productivity growth rate during Period III. But none of the growth rates of rice productivity in any of the periods studied was found statistically significant.

Wheat is the only crop which recorded increasing productivity growth rate in North Bihar throughout the period under

Table 5.11

Zone-wise Annual Growth Rate in Productivity of Principal Crops of North Bihar During Study Period and Sub-Periods

(in per cent)

Crop Categories	*Study Period (1970-71 to 1993-94)*	*Sub-Periods*		
		Period-I (1970-71 to 1979-80)	*Period-II (1980-81 to 1989-90)*	*Period-III (1990-91 to 1993-94)*
North Bihar				
Rice	1.3	3.8	4.0	-4.3
Wheat	1.5	3.4	2.0	1.2
Maize	3.8*	1.1*	8.5*	1.9
Barley	2.2**	-1.4	2.7	1.4
Arhar	2.3	1.7	-0.1	-7.7
Gram	2.1	2.2	-2.0	-5.6
Lentil	2.4	1.2*	2.2	-6.9**
Sugarcane	1.7	-1.1	7.3	0.7
Rape and mustard	2.7	0.1	1.9	5.9
Zone I				
Rice	1.5	3.3	3.5	-4.7
Wheat	1.5	4.0	2.6	2.3
Maize	5.6	9.6	8.7*	2.6
Barley	2.2***	-1.4	3.9	1.6
Arhar	2.6**	-0.1	-0.5	-6.2
Gram	2.1	3.8	-1.9	-0.5
Lentil	2.8	1.1**	3.5	-8.4
Sugarcane	1.7	-1.1	7.6	0.7
Rape and Mustard	3.1	0.5	2.1	9.6
Zone II				
Rice	1.2	5.5	4.6	-3.2
Wheat	2.1	-0.7	-0.2	-2.7
Maize	3.9**	6.6	8.5	0.2

Contd...

Contd...

Barley	1.4	-0.3*	1.4	-3.4
Arhar	2.6**	-3.4*	5.1	-11.3
Gram	1.1	-1.7	-3.1	-12.9
Lentil	0.08	1.6	-1.6	2.8
Sugarcane	1.5	-1.5	4.0*	3.0
Rape and Mustard	1.9	-0.08	0.09	0.2

* Significant at 10 per cent level of probability.
** Significant at 5 per cent level of probability.
*** Significant at 1 per cent level of probability.

study and over the sub-periods. Annual growth rate in wheat productivity was 1.5 per cent during study period in North Bihar, however, sub-period-wise analysis indicated increasing growth rate in wheat productivity. But the annual growth rate in wheat productivity was comparatively higher during Period I (3.4%) than the Period II (2%) and Period III (1.2%). Almost a similar trend was observed in Zone I. Zone II, no doubt, showed increasing growth rate in wheat productivity (2.1%) during the study period but the negative growth rates were observed in all the sub-periods, probably due to comparatively higher base in different sub-periods. Moreover, per ha productivity of wheat increased from 1,179 kg during Period I to 1,310 kg during Period II and 1,945 kg during Period III in Zone II.

In North Bihar, annual productivity growth rate of maize witnessed not only increasing but the growth rates obtained during the study period, Period I and Period II was statistically significant at 10 per cent level of probability. It was also around 2 per cent during Period III but it was not statistically significant. Zone I and Zone II also witnessed increasing growth rate in maize productivity during study period and all the sub-periods under study, but the growth rates obtained during Period II in Zone I and the study period in Zone II were found statistically significant.

An increasing growth rate in productivity of wheat and maize, during post-green revolution period can only be explained by increasing adoption of high yielding varieties of these crops and introduction of winter maize in the region.

Despite the significant decline in the area under barley, its annual productivity growth rate increased by 2.2 per cent in North Bihar during the study period which was statistically significant at 5 per cent level of probability. Annual productivity of barley showed negative growth during Period I (-1.4%), but a positive trend during Period II and Period III (2.7% and 1.4%). Zone I witnessed the similar productivity growth rates of barley as observed in North Bihar. Zone II also recorded an increasing trend in annual growth rate of barley productivity during the period under study but statistically significant negative rate of productivity during Period I and non-significant positive growth rate during Period II which turned to be negative during Period III.

All the principal pulse crops like *arhar*, gram and lentil registered increasing growth rate in their respective productivity during the period under study, not only in North Bihar but in both the zones under study. Zone I and Zone II witnessed identical growth rates in *arhar* productivity which was also statistically significant at 5 per cent level of probability. During the study period, growth rates in gram and lentil productivity were comparatively higher in Zone I than that of Zone II. Lentil is the only crop which registered statistically significant growth in productivity in North Bihar during Period I which increased during Period II (2.2%) but turned to be negative and statistically significant during Period III. A similar trend was observed in Zone I. Zone II had a quite different trend in lentil productivity as observed in Zone I. The lentil productivity declined from Period I to Period II and it registered growth of 2.8 per cent during Period III when negative growth was observed in Zone I.

The productivity growth rates of rape and mustard were 2.7 per cent, 3.1 per cent and 1.9 per cent in North Bihar, Zone I and Zone II respectively during the study period. However, the growth in productivity during sub-periods in North Bihar showed an increasing trend, i.e., 01 per cent, 1.9 per cent and 5.9 per cent during Period I, Period II and Period III, respectively. A similar trend was observed in Zone I and Zone II, i.e., 0.5 per cent during Period I, 2.1 per cent during Period II and 9.6 per cent during Period III in Zone I and -0.08 per cent during Period I, 0.09 per cent during Period II and 0.2% during Period III in Zone II.

Sugarcane productivity growth rates were also positive in North Bihar, Zone I and Zone II during the study period and it was identical in North Bihar and Zone I (1.7%) and comparatively lower in Zone II (1.5%). During the early stage of green revolution (i.e., in Period I) sugarcane productivity registered a negative growth rates in North Bihar (-1.1%) and Zone II (-1.5%) but positive growth rate in Zone I (1.1%). But the sugarcane showed a comparatively higher rate of productivity growth in North Bihar (7.3%) and Zone I (7.6%) and Zone II (4.0%) during Period II however, it was statistically significant only in Zone II. During Period III, there was a deceleration in productivity growth rate of sugarcane in North Bihar and both the zones under study.

Growth in Production

In North Bihar, annual production growth rate of rice was estimated to 0.2 per cent during the study period (Figure 2). Sub-period-wise analysis revealed that the production growth rate was less than 1 per cent (0.55%) during Period I which turned to be negative (-0.83%) during Period II and further declined to -4.3 per cent during Period III. Zone II registered a comparatively lower growth rate in rice production (1.6%) than that of Zone I (1.8%) during study period but the performance of Zone II had comparatively been better when sub-period-wise analysis was undertaken. Zone II recorded comparatively higher growth rate in rice production than that of Zone I during all the sub-periods under study. Zone I also witnessed negative growth in rice production during Period III. The decline in production growth rate in rice was mainly contributed by a decline in productivity, particularly during Period III, influenced by erratic behaviour of monsoon and network of rivers which caused longer inundation of standing paddy crop during the period in the project area. It is worth pointing out that the Zone II could maintain the positive growth rate in rice production due to an increase in productivity and area during Period II and Period III respectively.

Like other parts of the country, annual growth rate in wheat production showed an increasing trend in North Bihar (3.8%) during the study period (Figure 2). However, zone-wise analysis indicated comparatively higher growth rate in wheat productivity during study period in Zone I (5.7%) than Zone II (4.5%). North Bihar registered much higher annual growth in wheat production during Period I, but experienced a deceleration in growth rate during Period II and Period III, probably due to comparatively higher base in subsequent periods. Zone I also showed a similar trend in wheat production during subsequent periods. On the other hand, annual productivity growth rate showed acceleration in consecutive sub-periods in Zone II because the Zone II might be late starter in adoption of modern technology, particularly in wheat production.

Annual growth rate in maize production showed an increasing trend in North Bihar during the study period (Figure 2) which was found statistically significant at 1 per cent level of probability. During sub-periods, the production growth rate of maize also showed an increasing trend but it was statistically significant during Period II only. Zone II witnessed a much higher growth rate in maize production which was 5.5 percent during the study period, 12.5 per cent during Period I, 8.1 per cent during Period II and 9.1 per cent during Period III. It was again due to late adoption of winter maize in Zone II. The Zone I recorded comparatively a lower growth rate than that of Zone II but the growth rate in maize production achieved during Period II was significant at 5 per cent level of probability in Zone I. It is worth pointing out that the increasing trend in maize production during the period under study including all the three sub-periods has been mainly due to increasing trend in productivity which was made possible due to increasing adoption of winter maize in the area. Zone II performed better than Zone I, particularly in maize production due to late adoption of winter maize cultivation.

In North Bihar, barley witnessed a declining growth rate (-9.2%) in production during the period under study. However, the decline in barley production was comparatively higher in Period I (-12%) than Period II (-11.1%) but production growth rate turned to positive during Period III (0.2%). The Zone I observed almost identical trend in barley production during period under study and all the sub-periods. The Zone II also followed almost a similar trend but it registered a negative growth in barley production during all the three sub-periods under study.

In North Bihar, *arhar* and gram registered declining growth in their production but lentil recorded a growth of 2 per cent during the study period (Figure III). During subsequent sub-periods, almost a declining growth rate in *arhar* production was observed, however, gram witnessed a positive growth rate in production during Period I, though, it was not statistically significant, however, it followed the declining trend as observed in *arhar* production during Period II and Period III. It is worth pointing out that lentil recorded a significant increase in production during Period I, probably due to significant increase in lentil productivity during the period. The trend was maintained during the Period II (3.3%) but showed a significant declining trend (-6%) during Period III, again due to a significant decline in its productivity during the period. Zone-wise analysis revealed that *arhar* showed a declining trend in Zone I during the study periods and negative growth rate during all the sub-periods. It is worth pointing out that despite a significant increase in *arhar* productivity in Zone I during period under study, its production showed a significant declining trend, mainly due to a considerable decline in area (-3.2%) under *arhar* in Zone I during the period under study. On the other hand, Zone II registered a significant growth in *arhar* production (0.5%) during the period under study and recorded a moderate growth during Period II (3.3%) and a much higher growth during Period III (27.4%), mainly due to spectacular growth in *arhar* productivity during Period II and area during Period III (43.5%). On the other hand, Zone I observed a declining trend in *arhar* production during Period I and Period II. Gram also showed a declining trend in Zone I during the study period (-1.6%), Period II (-5.3%) and Period III (-3.2%). Annual growth rate in gram production showed a sign of recovery during Period I in Zone I and Zone II, but it was not statistically significant. The growth rate in gram production turned to negative in Zone I and Zone II during Period II but it showed increasing trend during Period III in latter zone and continued to be negative in the former.

Zone II performed comparatively better than Zone I in lentil production during the period under study and over the sub-periods under consideration. Zone I, no doubt, also showed an increasing trend in lentil production during the period under study (2.1%) and showed acceleration in growth rates during Period I and Period II which turned negative during Period III, mainly, due to decline in lentil productivity in the Zone (-8.4%).

Table 5.12
Zone-wise Annual Growth Rate in Production of Principal Crops of North Bihar During Study Period and Sub-Periods

(in per cent)

Crop Categories	Study Period (1970-71 to 1993-94)	Sub-Periods		
		Period-I (1970-71 to 1979-80)	Period-II (1980-81 to 1989-90)	Period-III (1990-91 to 1993-94)
North Bihar				
Rice	1.7	5.5	4.4	-4.3
Wheat	3.8	9.3	5.0	5.7
Maize	3.1***	4.5	5.8**	4.0
Barley	-9.2**	12.0	11.1	0.2
Arhar	-0.7	-2.6	-4.6	-3.4
Gram	-1.3	0.08	-0.6	-3.4
Lentil	2.2	1.7**	3.3	-6.0**
Sugarcane	1.8	-2.5	6.0	-0.7
Rape and Mustard	3.2	2.4	1.6	3.0
Zone I				
Rice	1.8	4.2	4.0	-6.5
Wheat	3.7	10.9	5.3	3.8*
Maize	2.5	2.5	5.3**	2.8
Barley	-9.1***	-12.6	-9.7	0.2
Arhar	-0.08*	-4.8	-3.7	-2.5
Gram	-1.6	0.08	-5.3	-3.2
Lentil	2.1	1.3	3.9	-8.6
Sugarcane	1.8	-2.7	6.3	-0.7
Rape and Mustard	3.9	2.7	2.4	4.5
Zone II				
Rice	1.6	8.7	4.0	0.08
Wheat	4.5	2.2	4.1	10.9
Maize	5.5	12.5	8.1	9.1

Contd...

Contd...

Barley	-10.1	-8.6*	-13.8	-3.8
Arhar	0.5*	-4.6**	3.3	27.4
Gram	-1.2	1.0	-9.2	7.6
Lentil	2.4	3.7	0.5	6.9
Sugarcane	-0.6	2.7	-11.4*	0.5
Rape and Mustard	1.7	2.3	-0.2	0.5

* Significant at 10 per cent level of probability.
** Significant at 5 per cent level of probability.
*** Significant at 1 per cent level of probability.

In North Bihar, the annual growth rate of rape and mustard production was estimated to 3.2 per cent during the study period which declined from 2.4 per cent in Period I to 1.6 per cent during Period II and again increased to 3 per cent during Period III. A similar trend was observed in Zone I also. But Zone II witnessed a negative growth rate in production of rape and mustard during Period II which was probably due to decline in area under rape and mustard during the period.

Sugarcane production growth rate was also positive (1.8%) in North Bihar during study period, but it was negative in Period I (-0.25%) and positive (6%) in Period II, however, it turned to negative (-0.7%) in Period III. Zone I, no doubt, had the identical growth rate in sugarcane production as observed in North Bihar during the study period, but recorded an increase in growth rate from Period I to Period II and again a similar negative growth rate as obtained in North Bihar during Period III. In Zone II, sugarcane production growth rate was negative (-0.6%) during study Period, however, it was positive in Period I and Period III but negative in Period II (-11.4%).

The increase in production growth rate of sugarcane has mainly been due to an increase in area under sugarcane. However, in Zone I, a significant increase in sugarcane production during Period II was contributed by increased productivity of sugarcane in the zone.

References

Mohammad, Ali, Khan, N. and Singh, T. (1979), "Regional variations in agricultural productivity-A study of Bihar", in *Dynamics of Agricultural Development* (ed.) Ali Mohammad, Delhi.

Singh, M. (1980), "Agricultural Development Differentials in States of Haryana and the Punjab: A Comparative Study", IARI, New Delhi, unpublished thesis, p. 125.

Sinha, D.K. (1991), "An Evaluation of Technological Progress in Bihar Agriculture with Special Reference to Rice: An Econometric Analysis", unpublished Ph.D. thesis, Department of Agricultural Economics, Rajendra Agricultural University, Pusa, Samastipur.

6

Instability in Agricultural Production

During the post-green revolution period, Indian agriculture witnessed a marked improvement, particularly in increasing production of foodgrain crops. There has been growth in the production of all the crops, though at different rate, but it has not been smooth for foodgrain crops in general and principal kharif crops in particular (Figure 5).

There are two factors which exerted influence on fluctuations in crop production, i.e., rainfall and input-output price ratio. Many studies have been conducted to identify the causes of fluctuations in foodgrain production, however, notable studies by Sen (1967), Mehra (1981), Hazell (1982), Ray (1983), Mahendra Dev (1987) and Hanumantha Rao *et al.* (1988) analysed the causes of fluctuations in foodgrain production in detail. These studies were conducted at all-India level and particularly for foodgrains as a group. There is a dearth of studies dealing with instability in growth of various crops at regional level. Hence, in the present section, an attempt has been made to study the instability in production of principal crops during sub-periods of green revolution in North Bihar.

FIGURE 5

Trends in Output Instability

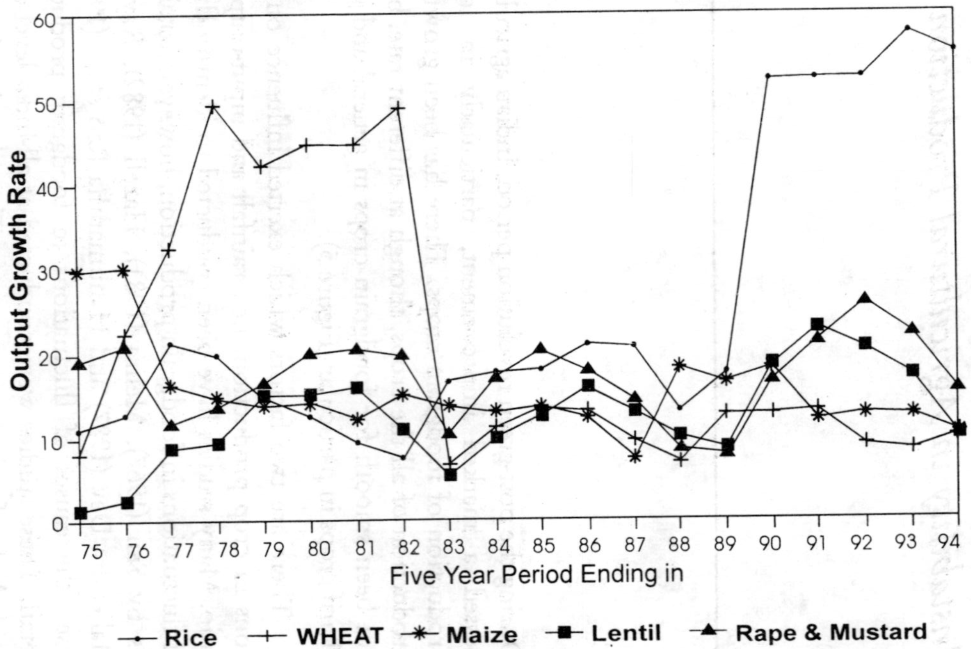

Since our objective is to study the fluctuation or deviation in output of principal crops grown, per cent annual change in output and standard deviation (SD) in annual changes were estimated to measure fluctuations in production of principal crops in North Bihar. The study covers the study period (1970-71 to 1993-94) and all the three specified sub-periods adopted while studying the growth in earlier chapter. At first, the instabilities in major crop categories like, foodgrains, cereals, pulses and oilseeds are examined. An attempt has also been made to study the fluctuations in major principal crops.

Table 6.1

Instability in Crop Production During Study Period and Sub-Periods in North Bihar

(in per cent)

Crop/Crop Group	*Study Period (1970-71 to 1993-94)*	*Sub-Periods*		
		Period-I (1970-71 to 1979-80)	*Period-II (1980-81 to 1989-90)*	*Period-III (1990-91 to 1993-94)*
Foodgrain	21.17	12.66	17.06	5.41
Total Cereals	20.22	11.14	16.32	6.61
Rice	29.20	19.61	36.91	27.52
Wheat	34.63	50.04	16.87	6.94
Maize	26.41	22.26	20.69	11.77
Barley	70.18	38.64	47.31	1.29
Total Pulses	16.84	22.18	11.29	10.10
Arhar	21.45	25.72	18.39	5.93
Gram	17.54	11.33	21.63	19.25
Lentil	20.29	11.10	17.77	6.02
Total Oilseeds	26.24	13.48	18.45	4.34
Rape and Mustard	39.23	19.21	22.46	14.8
Sugarcane	28.18	14.60	25.40	18.62

The perusal of Table 6.1 indicated that the SD of foodgrain production was 31.17 per cent during the study period in North Bihar, whereas the fluctuations in cereal and pulse production were 20.22 per cent and 16.84 per cent respectively. On the other hand, fluctuation in oilseed production was estimated to 26.24 per cent, indicating comparatively higher fluctuation in production of this category of crops than fluctuations observed in other crop categories during the study period. It was expected also since rape and mustard—major oilseed crop—is prone to pests and diseases which became more prominent in cultivation of local non-descript varieties of this particular crop. Major cereals like rice, wheat, maize and barley showed comparatively higher fluctuation in production than the fluctuation in cereals as a whole. The phenomenon may be explained by the compensating effect of various cereal crops in minimising the fluctuation in cereals as a whole. Among the different crop categories, pulses showed declining instability during consecutive sub-periods, whereas foodgrains, cereals, and oilseeds witnessed an increase in instability from Period I to Period II but became more stable during Period III. Among principal foodgrains, wheat, maize and *arhar* witnessed a continuous decline in instability over the three sub-periods. But the rice—principal crop of the project area—witnessed a sharp increase in instability from Period I to Period II (19.61 per cent to 36.91 per cent), whereas instability declined during Period III (27.52 per cent). Other crops followed a similar trend, i.e., increase in instability during Period II and decline during Period III. It is worth pointing out that wheat had the highest instability during Period I (50.04%) which declined to almost at the lowest level (6.94%) during Period III. It is also worth pointing out that all the crops under study witnessed a decline in the production instability during early nineties.

An attempt has also been made to study the comparative instability in principal crops grown in Zone I and Zone II during period under study and specified sub-periods. The estimated zone-wise instability in production of principal crops during specified periods is presented in Table 6.2.

Table 6.2

Instability in Crop Production During Study Period and Sub-Periods in Zone I and Zone II

(in per cent)

Crop/Crop Group	*Study Period (1970-71 to 1993-94)*	*Sub-Periods*		
		Period-I (1970-71 to 1979-80)	*Period-II (1980-81 to 1989-90)*	*Period-III (1990-91 to 1993-94)*
Zone I				
Total Foodgrains	21.07	10.63	16.88	1.84
Total Cereals	19.80	10.31	15.92	6.40
Rice	24.25	19.48	19.72	33.07
Wheat	36.91	57.72	17.87	5.40
Maize	24.33	22.37	19.40	60.45
Barley	72.90	41.03	31.43	1.27
Total Pulses	30.04	48.00	12.58	4.87
Arhar	21.20	25.01	15.59	8.52
Gram	20.74	13.55	23.89	25.02
Lentil	21.52	9.88	19.71	16.13
Total Oilseeds	34.06	16.38	28.29	6.10
Rape and Mustard	47.13	28.58	29.36	17.94
Sugarcane	28.31	14.90	25.84	18.83
Zone II				
Total Foodgrains	24.92	23.42	18.38	21.81
Total Cereals	22.78	16.67	19.87	22.95
Rice	25.50	29.23	22.08	23.98
Wheat	39.93	18.47	18.47	15.50

Contd...

Contd...

Maize	42.71	40.73	27.18	28.42
Barley	69.90	30.05	55.74	9.51
Total Pulses	14.39	11.66	12.11	22.42
Arhar	34.32	40.04	30.43	36.43
Gram	45.04	22.46	46.35	79.28
Lentil	22.62	21.41	15.99	11.81
Total Oilseeds	18.48	11.39	17.01	11.94
Rape and Mustard	30.75	17.58	17.14	12.35
Sugarcane	45.76	25.91	53.46	13.43

The instability in foodgrain production was comparatively higher in Zone II (24.92%) than Zone I (21.07%). However, Zone I witnessed an increase in instability of foodgrain production from Period I to Period II but it attained almost stability during Period III (1.84%). On the other hand, Zone II recorded a decline in instability from Period I to Period II which increased during Period III (21.81%). It clearly indicated that Zone I achieved almost stability in foodgrain production, whereas Zone II experienced a considerable fluctuation in foodgrain production. Wheat, barley and *arhar* showed declining instability in Zone I, whereas, wheat, lentil and rape and mustard showed declining instability in Zone II during three specified sub-periods under study. Rice showed increasing instability in Zone I but declining in Zone II, except a marginal increase during Period III. Instability in maize production was, no doubt, comparatively higher in Zone II than that of Zone I, but the former showed almost a declining trend. Despite the comparatively lower instability in maize production, Zone I experienced much higher level of instability (60.45%) during Period III. Barley is less important crop with declining area in both the zones but its production has almost been stabilised in Zone I, particularly during Period III, whereas its production was more unstable in Zone II during Period I and Period II, but got stabilised during Period III. Pulses are the most sensitive crops but showed declining instability in Zone I and increasing instability in Zone II. As explained earlier, *arhar* showed declining instability in Zone I and lentil in Zone II.

On the basis of above discussion, it may be said that wheat is the only crop which showed declining instability and an increasing growth in production during period under study as well as during different sub-periods. The production of various crops seemed to be stabilised during the period 1990-94 as compared to the period 1981-90. However, a clear-cut trend is yet to be emerged.

Instability in Area and Productivity

Instability in production of principal crops is expected to be caused by instability in area and productivity. If the instability in both the components declined, the instability in production has to be declined during particular period of time. To test this hypothesis, SD of area and productivity of principal crops were computed and have been presented in Table 6.3.

It is interesting to observe that instability in area and productivity in most of the crops fluctuated in the same direction, i.e., if there is an increase/decrease in instability in the area of particular crop, the instability in productivity also increases/decreases. It has been observed that the instability in area and productivity of wheat, maize, lentil and rape and mustard declined continuously during all the sub-periods under study (Table 6.3). On the other hand, instability in area and productivity of rice increased simultaneously from Period I to Period II and declined during Period III. *Arhar* and gram are the crops which experienced a declining trend of instability in area but yield did not show any specific trend.

Table 6.3
Instability in Area and Productivity of Principal Crops During Three Sub-Periods in North Bihar

(in per cent)

Crop		*Period-I*	*Period-II*	*Period-III*
Rice				
	A	7.21	11.54	8.01
	Y	15.52	35.80	21.70
Wheat				
	A	19.01	10.47	6.17
	Y	37.85	7.88	2.55
Maize				
	A	66.72	11.95	4.37
	Y	90.03	29.21	8.35
Barley				
	A	33.43	40.90	5.22
	Y	13.38	31.37	6.25
Arhar				
	A	22.04	16.05	10.17
	Y	21.42	5.59	16.65
Gram				
	A	5.75	15.08	10.60
	Y	14.42	15.41	21.37
Lentil				
	A	46.72	9.04	5.10
	Y	25.00	12.70	9.21
Rape and Mustard				
	A	31.02	6.74	6.40
	Y	26.98	17.82	15.58
Sugarcane				
	A	7.86	12.45	1.55
	Y	10.63	41.73	18.02

A = area; Y = yield

On the basis of trends in area and productivity of principal crops in North Bihar during different sub-periods under study, it may be said that the instability in production of principal crops is jointly influenced by area and productivity which generally move in the same direction, but it could not be ascertained in the present study as to which of the components (area or productivity) was the causal factor, however, their respective contribution to production instability have been estimated.

An explanation for increased instability in foodgrain production may be found in the nature of environmental and infrastructural conditions under which growth is brought about[1]. But the present study does not subscribe the above view. Despite the improvement in infrastructural facilities, the instability in rice and maize production increased during recent past (Period III), whereas the instability in wheat production declined during the period in North Bihar in general, and Zone I and Zone II in particular.

As discussed earlier, the instability in area and yield generally move in same direction, but area instability is generally lower than the yield instability for almost all the principal crops except wheat (Table 6.3). Hence, the expansion of the former must have had a stabilising impact on output fluctuations. However, in the process of expansion, marginal and less fertile lands might have been increasingly brought under cultivation. Since yields from such marginal land would be more sensitive to variations in rainfall, this might have contributed to increased instability in yields of many crops.

The variability in annual output can be decomposed into: (a) variability in area, (b) variability in yield, and (c) synchronous movements in area and yield. Table 6.4 reveals that the yield variability has been the major factor (89.58%) for instability in rice production in North Bihar during the period under study. The contribution of correlated factor (area and yield) was negative and substantial in rice production indicating the stabilising impact of this interaction. Yield variabilities were also causing comparatively more instability in wheat, maize and rape and mustard than the instability caused by respective area variability. The variability in area has caused comparatively more instability in only lentil production. The interaction of area and yield have caused a considerable influence on instability of production of wheat, maize, lentil and rape and mustard.

Table 6.4

Decomposition of Instability in Annual Crop Output Growth Rate During Study Period (1970-71 to 1993-94) in North Bihar

(in per cent)

Crops	*Area*	*Yield*	*Interaction*
Rice	34.22	89.58	-23.80
Wheat	16.95	24.51	58.52
Maize	8.52	8.91	82.56
Lentil	19.90	12.87	67.21
Rape and Mustard	19.25	21.82	58.91

Hence, it may be said that the yield induced instability was more pronounced in crops which got more attention by biological scientists for varietal improvement and new package of practices. These crops are generally grown in irrigated area. The yield induced production instability was comparatively lower in lentil since a very little has been done to harness its genetic potential and farmers are still using the local seeds and traditional package of practices for cultivation of this crop.

In almost all the crops under study, area-yield interaction induced instability was much higher indicating that the increase in area variability is accompanied with the increase in yield variability. It may be due to uncertain agriculture in the area due to poor irrigation facilities.

Sensitivity of Output to Rainfall Variation

Various studies suggest that the variation in rainfall has caused the fluctuations in production of foodgrain crops during recent years. The fluctuation in crop output has also been found to be the combined result of the sensitivity of the output to variation in rainfall and variability in weather itself (Ray, 1983 and Hanumantha Rao *et al.*, 1988). The variability of rainfall over a shorter period of time may be high or low, but over a longer period one would not expect a marked change in its variability. However, the sensitivity of output to variation in rainfall may change depending upon nature of technology and inputs used and also upon the level of development in infrastructural facilities like power, irrigation and agriculture extension and input supply system. In general, the sensitivity of output to rainfall variation is found to be higher in low rainfall area and lower as the level of rainfall rises. Moreover, the moisture is an important input and an increase in moisture due to high rainfall may result in diminishing return, particularly in crops which do not thrive well in high moisture conditions.

Table 6.5 gives the percentage deviation in production due to 1 per cent deviation of rainfall during the period under study. The production of wheat and rice---the major foodgrain crops in the area constituting nearly 75 per cent of foodgrain production---had positive influence of rainfall. An increase of 1 per cent of rainfall would contribute 0.08 per cent in rice production and 0.07 per cent in wheat production in the project area. Maize, still having larger area in kharif season, is found more sensitive to rainfall, hence, the 1 per cent increase or decrease in rainfall would influence the production 0.22 per cent in opposite direction. Among the rabi crops, gram and rape and mustard

Table 6.5
Percentage Deviation in Production due to 1 per cent Deviation in Rainfall

Crops	*Value of Regression Coefficient (b)*
Rice	0.08
Wheat	0.065
Maize	-0.22
Barley	-0.16
Arhar	-0.12
Gram	-6.36
Lentil	-0.67
Rape and Mustard	-1.04
Sugarcane	negli.*

* negli. = negligible.

have been found to be more sensitive to rainfall since increase in one per cent of rainfall would reduce the production of gram by 6.36 per cent and rape and mustard by 1.04 per cent. It may be probably due to the fact that high level of rainfall during September is responsible for delay in sowing of these crops and winter rains create congenial environment for diseases and pests in these crops. The output of *arhar* is less sensitive to rainfall in comparison to other pulses because it can be sown during longer period of time and standing crop is comparatively less influenced adversely due to winter rain.

Growth and Instability

In the present sub-section, the relationship between growth rate and instability in major principal crops grown in the project area has been examined. An effort has also been made to assess the performance of these crops by making matrix, based on rate of growth and instability. The regressions between growth and

Table 6.6

Annual Growth Rate and Standard Deviation in Production of Principal Crops in North Bihar During Sub-Periods

(in per cent)

Crop	*Period I (1970-71 to 1979-80)*	*Period II (1980-81 to 1989-90)*	*Period III (1990-91 to 1993-94*
Rice			
Growth Rate	5.5	4.0	-4.3
SD	19.61	36.91	27.52
Wheat			
Growth Rate	9.3	5.0	5.7
SD	50.04	16.87	6.94
Maize			
Growth Rate	4.5	5.8**	6.94
SD	22.26	20.69	11.77
Barley			
Growth Rate	-12.0	-11.1	0.2
SD	38.64	47.31	1.29
Gram			
Growth Rate	0.08	-0.6	-3.4
SD	11.33	21.63	19.25
Arhar			
Growth Rate	-2.6	-4.6	-3.6
SD	25.72	18.39	5.93
Lentil			
Growth Rate	1.7**	3.3	-6.9**
SD	11.10	17.77	6.02
Rape and Mustard			
Growth Rate	-2.5	6.0	-0.7
SD	19.21	22.46	14.80
Sugarcane			
Growth Rate	2.4	1.6	3.0
SD	14.60	25.40	18.62

its instability for principal crops were also estimated to get reliable information about their relationships during period under study. The growth rate and instability (SD) of principal crops are presented in Table 6.6.

In North Bihar, there was a deceleration in growth rate of rice production but no specific trend was observed in its instability during sub-periods under study. Despite the decline in growth rate in rice production from 5.5 per cent during Period I to 4 per cent during Period II, the respective instability in rice production increased from 19.61 per cent to 36.91 per cent, the growth rate further declined and instability followed the same trend from Period II to Period III. Growth rate in wheat production declined from Period I to Period II and again increased in Period III but instability in wheat production declined continuously from 50.04 per cent in Period I to 6.9 per cent in Period III. In North Bihar, maize was the only cereal crop which achieved a significant growth in production (5.8%) during Period II and had the growth rate of 4.5 per cent during Period I and 4 per cent during Period III, but the instability showed a declining trend over sub-periods.

Among the crops under study, barley registered comparatively more decline in production, i.e., -12.00 per cent in Period I and -11.1 per cent during Period II but its respective instability in production increased during the period, i.e., from 38.64 per cent to 47.31 per cent, respectively. The crop achieved the positive growth rate of 0.2 per cent during Period III, but its instability in production declined to 1.29 per cent.

Zone-wise analysis for examining the relationship between growth and instability has also been under taken to get the information on this particular issue in smaller area since the compensating effect of one area to another area may dilute the relationship between growth and instability of a particular crop. The relevant data are presented in Tables 6.7 and 6.8.

Table 6.7

Annual Growth Rate and Standard Deviation in Production of Principal Crops in Zone I During Sub-Periods

(in per cent)

Crop	*Period I (1970-71 to 1979-80)*	*Period II (1980-81 to 1989-90)*	*Period III (1990-91 to 1993-94*
Rice			
Growth Rate	4.2	4.0	-6.5
SD	10.48	19.72	33.07
Wheat			
Growth Rate	10.9	5.3	3.8
SD	57.72	17.87	5.40
Maize			
Growth Rate	2.5	5.3	2.8
SD	22.37	19.40	60.45
Barley			
Growth Rate	-12.6	-9.7	0.2
SD	41.03	31.43	1.27
Arhar			
Growth Rate	-4.8	-3.7	-2.5
SD	25.01	15.59	8.52
Gram			
Growth Rate	0.08	-5.3	-3.2
SD	13.55	23.89	25.02
Lentil			
Growth Rate	1.3	3.9	-8.6
SD	9.88	19.71	16.13
Rape & Mustard			
Growth Rate	2.7	2.4	4.5
SD	28.58	29.36	17.94
Sugarcane			
Growth Rate	2.7	6.3	-0.7
SD	14.90	25.84	18.83

Table 6.8

Annual Growth Rate and Standard Deviation in Production of Principal Crops in Zone II During Sub-Periods

(in per cent)

Crop	*Period I (1970-71 to 1979-80)*	*Period II (1980-81 to 1989-90)*	*Period III (1990-91 to 1993-94*
Rice			
Growth Rate	8.7	4.0	0.08
SD	29.23	22.08	23.98
Wheat			
Growth Rate	2.2	4.1	10.9
SD	18.47	18.47	15.50
Maize			
Growth Rate	12.5	8.1	9.1
SD	40.73	27.18	28.42
Barley			
Growth Rate	-8.6	13.8	-3.8
SD	30.05	55.74	9.51
Arhar			
Growth Rate	-4.6	-3.3	27.4
SD	40.04	30.43	36.43
Gram			
Growth Rate	1.0	-9.2	7.6
SD	22.46	46.35	79.28
Lentil			
Growth Rate	3.7	0.5	6.9
SD	21.41	15.99	11.81
Rape and Mustard			
Growth Rate	2.3	-0.2	0.5
SD	17.58	17.14	12.35
Sugarcane			
Growth Rate	2.7	-11.4	0.5
SD	25.91	53.46	13.43

A perusal of the tables indicates that there was a deceleration in the growth rate in rice production during three sub-periods under study in both the zones, but instability showed an acceleration in Zone I, whereas no clear-cut trend in production instability emerged in Zone II. On the other hand, growth rates in wheat production also showed deceleration and instability also followed the same trend throughout the different sub-periods under study. Maize did not show any specific trend in rate of growth and instability, but there was the highest instability in maize production during Period III (60.45%) whereas, the growth rate was only 2.8 per cent. The above analysis did not show any specific relationship between growth and instability which is also true for other crops under study. In Zone II also, positive relationship of growth and instability has not been observed in any of the crops under study. Hence, the tabular analysis failed to establish any relationship between growth and instability.

Classification of Crops on the Basis of Rates in Growth and Instability

In order to assess the relationship between growth and instability in production of principal crops, a matrix of crops on the basis of growth and instability was prepared. The crops are cross classified on the basis of rate of growth and level of instability recorded during the study period 1970-71 to 1993-94 as shown in Table 6.9.

Among the crops, wheat and maize achieved the growth rate of more than 3 per cent but instability in wheat production was more than 30 per cent, and 26.41 per cent for maize production. Barley and rape and mustard registered the lowest growth rates in production (less than 1.5%) but the instabilities in production of both the crops were more than 30 per cent during period under study. *Arhar* and gram also registered a growth rate of less than 1.5 per cent but the instabilities in their productions were less than 25 per cent.

Table 6.9

Cross Classification of Crops According to Level of Instability and Growth Rate in Production Achieved During Study Period in North Bihar

(in per cent)

Instability	*Annual Growth Rate*		
	3% and above	*1.5% to 3%*	*Less than 1.5%*
25%	-	Lentil	Gram, *Arhar*
25% to 30%	Maize	Sugarcane	Rice
30% and above	Wheat	-	Barley, Rape and Mustard

It may be worth mentioning that the growth is necessary condition for the success of agriculture in any region. Low instability without growth does not have any meaning. The declining instability along with increasing growth in agricultural production would be an ideal situation which has been observed in case of wheat production only. However, it could only be reliably said after analysing regression estimates of growth and instability of different crops under study.

Regression estimates between moving growth rate and instability (SD) were estimated for principal crops such as rice, wheat, maize, lentil, and rape and mustard. It was estimated for three periods, i.e., study period (1970-75 to 1989-94), early period of green revolution (1970-75 to 1979-84) and latter period of green revolution (1980-85 to 1989-94). The regression estimates are presented in Table 6.10.

Table 6.10

Regression Estimates Between Growth Rates and Instabilities of Principal Crops During Study Period, Early Period and Latter Period of Green Revolution

Crop	*Study Period*	*Early Period of Green Revolution (1970-75 to 79-89)*	*Latter Period of Green Revolution (1980-85 to 89-94)*
Rice	0.005	0.503	-0.032
Wheat	-0.94932	-0.14134	0.69768**
Maize	0.756**	0.924**	0.954
Lentil	0.70433***	0.52813***	0.99947*
Rape and Mustard	0.45770	0.86769	-0.49966***

During the study period, there was positive relationship between growth and instability in rice production. During sub-periods, the relationship was found positive during early period of green revolution and negative during latter period of green revolution. However, the relationship was not found statistically significant in either of the periods under study. In wheat, the relationship was negative during study period and early period of green revolution but it turned to be positive and statistically significant during the latter period of green revolution. On the other hand, the relationship in case of maize was positive and significant during study period and early period of green revolution. It remained positive during latter period of green revolution but it was not statistically significant.

Lentil is the only crop which had positive and significant relationship between its growth and instability in production during all the three periods under considerations. Rape and mustard had positive but non-significant relationship during study period and early period of green revolution which turned to be negative and statistically significant during latter period of green revolution.

It is very difficult to find out any clear-cut conclusion from the above discussion. Rice, which is grown in diversified situation, had no specific relationship between growth and instability in production. It is worth pointing out that the crops, irrespective of the level of technological improvement gained, had the positive relationship between growth and instability in production. Rape and mustard which lost area during the period under study had negative relationship between growth and instability. It happened due to bringing marginal land under cropping which helped increasing growth in production and instability. Hence, it may be said that productivity induced growth would be more stabilised than area induced growth in production.

Note

1. Hanumantha Rao, C.H., Ray, S.K. and Subbarao, K. (1988), op. cit. p. 35.

References

Hanumantha Rao, C.H., Ray, S.K. and Subbarao, K. (1988) "Unstable Agriculture and Droughts", New Delhi. Vikas Publishing House Pvt. Ltd.

Hazell, Peter B.R. (1982), "Instability in Indian Foodgrain Production International Food Policy Research Institute," *Research Report* No.30, Washington, D.C.

Mahendra Dev, S. (1987). "Growth and Instability in Foodgrains Production", *Economic and Political Weekly* 22 (39): A-82.

Mehra, S. (1981), "Instability in Indian Agriculture in the Context of New Technology International Food Policy Research Institute," *Research Report* No. 25, Washington, D.C.

Ray, S.K. (1983), "An Empirical Investigation of Nature and Causes for Growth and Instability in Indian Agriculture", *Indian Journal of Agricultural Economics* 38(4): 459.

Sen, S.R. (1967), "Growth and Instability in Indian Agriculture", *Agricultural Situation in India*, 21(10): 827.

7

Components of Foodgrain Production/Growth

In previous chapters, cropping pattern, changes in area, production and productivity of principal crops, instability in their productions and relationship between growth and instability were analysed and discussed. But these analyses may not convey the complete scenario of growth and instability unless a detailed analysis on components of foodgrain production and growth is undertaken. Keeping in view the importance of production and growth promoting components, the quantitative assessment of various factors in agricultural production/growth in North Bihar was planned to generate information which would be helpful in reorienting the programmes and priorities of agriculture development to achieve the higher rates of growth. The factors like gross area sown, irrigated area, fertilizer consumption, rainfall, coverage under HYVs, land development, soil conservation, package of practices, extent of mechanisation, improved implements, input supply system, agricultural extension services, agricultural research, administered princes, and marketing structure may be included for generating precise information on components of production/growth. But the introduction of multiplicity of factors in any mathematical model to identify and evaluate their contribution to the production/growth of agricultural output is, however, limited by the absence of comparable time series data on many of the factors. The limited number of years for which time series data were available in desired format also restricts the number of factors that could be included in the analysis unless the suitable transformation is applied. Moreover, quantification of qualitative factors like, agricultural extension services, agricultural research and package of practices of different crops also restrict the inclusion of these variables in the statistical model.

Many attempts were made earlier to study the components of production/growth in agriculture production, but most of them are conducted at national level, mainly encompassing the post-independence period, i.e., pre and post-green revolution period. A state-wise analysis covering the states of Bihar, Kerala, Madhya Pradesh, Tamil Nadu, Karnataka, Rajasthan, Uttar Pradesh and West Bengal was undertaken to measure the contribution to growth in crop output (Parikh, 1966). The coefficient of land was found positive and significant in all the states except Tamil Nadu. The growth of crop output was explained by land to the extent of 70 per cent or more in more than half of states and 20 to 50 per cent in remaining states. The coefficient of irrigation was positive in Tamil Nadu, Bihar, Kerala and Uttar Pradesh but it was significant in Tamil Nadu only. The contribution of fertilizer was 49 per cent in Kerala, 31 per cent in Tamil Nadu, 14 per cent in Bihar and less than 8 per cent in all the other states.

Giri *et al.* (1966) also studied the components of crop output growth in India by analysing time series data on gross area sown, irrigated area and consumption of fertilizer in pre-green revolution period and found that the expansion of area sown contributed about three-fourth of the growth in crop output in

India, during 1951-52 to 1962-63. Fertilizer played much less important role in increasing the growth in crop output since it contributed less than 20 per cent to the growth in agriculture production during the period under study. The irrigation failed to record any contribution to growth in crop output in pre-green revolution period.

Kumar and Kaul (1991) also studied the contribution of growth components and instability in agriculture during pre- and post-green revolution periods. During the pre-green revolution period, they found land as a major component of crop output growth which remained an important component in the post-green revolution period (1966-82) also. During the post-green revolution period, fertilizer emerged as an important input of crop output growth (18%) whereas HYV and irrigated area could only register a marginal contribution of 5 per cent each to growth in crop output during the post-green revolution period. Among the different independent factors of production considered for such type of analysis, rainfall has been found exerting significant influence on crop production at all-India level in general and state of Bihar in particular (Cummings and Ray, 1969; Ray, 1979 and Hanumantha Rao, C.H., 1988). Keeping in view the findings of these studies, the regression estimates of rainfall to production of principal crops of North Bihar were estimated for the post-green revolution period. Rainfall regression coefficient for none of the crop was found statistically significant. Moreover, at the present stage of development, the impact of each of these factors taken separately may result in small regression estimate like rainfall but their combined effect on production would give some meaningful result (Giri *et al.*, 1966).

These studies adopted production function of Cob-Duglus type and used data of different variables by either making index or transformed the data in desired proportions. Such an analysis covering a heterogeneous agricultural situation in a vast country like India necessitates this type of transformation to reduce/eliminate the multicollinearity among independent variables. Despite the preciseness in analysis, the findings of studies conducted at national level may not be valid for region having a typical agro-climatic situation like North Bihar. In the present analysis, the usually employed production function of Cob-Duglus (double lcg) type was fitted to the time series data from 1970-71 to 1993-94. The annual foodgrain production was considered as dependent variable and gross cropped area, irrigated area, fertilizer consumption, coverage under HYVs of seeds and annual rainfall as independent variables. The model does not take into account other factors either due to non-availability of time series data or due to problem in quantification.

At first stage, model was fitted to zone-wise data and the results there of are presented in Table 7.1. For this analysis also, the dependent variable considered was foodgrain production(Y) and the explanatory variables were gross cropped area (X_1), gross irrigated area (X_2) fertilizer consumption (X_3), HYVs are (X_4) and annual rainfall (X_5).

In both the zones, all the five variables included in the model explained only 57 percent variation in foodgrain production. In Zone I, land, irrigation and rainfall were found exerting a significant influence on foodgrain production but the regression coefficients of fertilizer and HYVs were negative but not statistically significant (Table 7.1). In Zone II, the regression coefficient of land was positive and statistically significant. The regression coefficients of irrigation, HYVs and rainfall were positive, and coefficient of fertilizer was found negative but none of these regression coefficients was statistically significant. Our purpose was to compare the performance of different factors in agricultural production/growth in two zones under study. The zone-wise analysis only indicated that land is still exerting a significant influence on agriculture production in both

Table 7.1

Regression Coefficients and Corresponding, Standard Error and Respective R^2 of Different Variables in North Bihar (Zone I-Zone II During Different Specified Periods Under Study)

Region	Regression Coefficient and Standard Error					
	Gross Cropped Area X_1	Gross irrigated Area X_2	Fertilizer Consumption X_3	HYV	Rainfall	R^2
Zone I	1.8422**	1.6323*	-2.3439	-32.22	1573.7**	0.573
	(0.83422)	(0.86604)	(2.3691)	(401.40)	(602.39),	
Zone II	0.60300***	0.30758	-1.0893	2.9050	121.44	0.577
	(0.17742)	(0.31787)	(0.99364)	(28.082)	(86.670),	
North Bihar (Study Period) (1970-71 to 93-94)	1.3309*	1.3158**	-1.0637	-402.35	1992.0***	0.677
	(0.64561)	(0.60921)	(1.7329)	(308.90)	(689.83),	
North Bihar (Period I) (1970-71 to 81-82)	0.07537	1.5566	10.754	-426.94	3095.6**	0.702
	(0.73156)	(0.77957)	(18.711)	(797.71)	(1097.9),	
North Bihar (Period II) (1982-83 to 93-94)	3.1740**	1.1838	0.88852	-2313.2	1120.9	0.817
	(1.1271)	(1.0934)	(3.1838)	(1490.0)	(1017.7),	
North Bihar with 4 Variables	0.94306	1.2059*	-	-1.9194	21.528***	0.646
	(0.58324)	(0.61430)		(1.6326)	(6.9101),	
North Bihar with 4 Variables	0.99969*	0.57473*		-	0.2183***	0.621
	(0.58677)	(0.30147)			(0.06971),	

Figures in parentheses are the standard error.

the zones, but irrigated area and rainfall influenced the agriculture production significantly in Zone I only and had non-significant positive influence in Zone II. The results can only be explained by excessive rainfall and water congestion in most of the area due to typical land situation and menace of flood in Zone II. In such a situation, it would be most difficult to find out any relation of independent variables with the dependent variables, particularly through time series data. At the second stage, the same model with the same independent and dependent variables was fitted to the data of North Bihar as a whole. The coefficient of multiple determination (R^2) was found comparatively higher (68%) indicating that the variables included in the model explain nearly 68 per cent of variation in foodgrain output during the period under study. Regression coefficients of land, irrigated area and rainfall were positive and significant, whereas regression coefficients of fertilizer, and HYVs were found negative indicating a significant influence of former variables in the production of foodgrain in the project area. The model was also fitted to North Bihar for the period 1970-71 to 1981-82 and the period 1982-83 to 1993-94 to find out the strength of dependent variables in increasing the agriculture production during two periods, i.e., early and latter periods of green revolution. Surprisingly, the regression coefficient of land was positive but not significant in the early period of green revolution but its regression coefficient was much higher (positive) and statistically significant during latter period of green revolution indicating that the factors not included in model like extension services, input supply system, credit, etc., might have played major role in foodgrain production in early period of green revolution but land emerged as an important factor in increasing agricultural production during latter period of green revolution. The regression coefficient of rainfall was much higher and statistically significant during early period of green revolution which remained positive but turned to be statistically non-significant during latter period of green revolution. The phenomena can only be explained through increased area under irrigation during latter period of green revolution which made the agriculture comparatively less dependent on rainfall. Moreover, it requires in-depth analysis. Regression coefficients of irrigated area and fertilizer were positive and of HYVs was negative in both the periods, but none of the these coefficients were statistically significant.

The main thrust of the present analysis was directed to the estimation of contribution of important factors of production in growth of foodgrain production in North Bihar. The negative and non-significant coefficients of fertilizer and HYVs, appeared in the model was probably due to the fact that effects of these variables might have been distributed among other variables. Besides, the increase in fertilizer consumption and area under HYVs has not been proved as a necessary condition for increasing the foodgrain production in the area. Despite the continuous increase in these two variables, the foodgrain production witnessed many troughs resulting into negative regression coefficients of these variables in the model under consideration. It is worth pointing out that these variables were found exerting significant influence on production of principal crops in the project area in various micro-analysis of cross-sectional data (Sharma, 1991).

It clearly indicated positive and significant influence of these variables on production of foodgrains in the project area. An increase of 1 per cent in gross cropped area, irrigated area and rainfall would likely to increase the foodgrain production by 0.99 per cent, 0.3 per cent and 0.21 per cent respectively in North Bihar. The R^2 was bifurcated to know the contribution of significant variables (gross cropped area, irrigated area and rainfall) in foodgrain production which are presented in Table 7.2.

Table 7.2
Contribution of Significant Variables in Foodgrain Production in North Bihar

Variables	*Contribution (%)*
Gross Cropped Area	20.91
Gross Irrigated Area	16.29
Annual Rainfall	24.90
Total contribution explained by these variables	62.1

It may be observed from the table that the rainfall is still in position to explain nearly 24.90 per cent of variation in foodgrain production in the project area. Gross cropped area has also been able to explain 20.91 per cent variation in foodgrain production, indicating importance of land in agricultural production. Irrigation, no doubt, is an important factor for agricultural production but it has explained only 16.29 per cent of variation in foodgrain production. It was only due to poor irrigation infrastructure in the project area. The study indicates that with the increase in irrigated area by 1 per cent in North Bihar, its contribution to the variation of foodgrain production has been at the lowest level of 16.29 per cent, probably due to inefficient canal system, non-functioning of public tubewells, erratic supply of electricity, and inadequate availability of diesel in the project area during the period under study.

Components of Growth

In the preceding subsection, an effort was made to find out the contribution of different variables in production of foodgrains in the project area. However, the present subsection deals with the importance of different factors of the growth of foodgrain production. The earlier analysis indicated the extent of influence of individual factor in production of foodgrains in the project area whereas the present analysis would reveal the contribution of these factors to the growth of foodgrain output during the period under study in the project area.

To estimate the contribution of these significant variables (GCA, irrigation and rainfall) to the growth in foodgrain production, the following model was adopted:

$$Gy = B_1Gx_1 + B_2Gx_2 + B_5Gx_5$$

where, Gy, Gx_1, Gx_2 and Gx_5 are the compound growth rates of foodgrain output, gross cropped area, irrigated area and rainfall respectively, whereas B_1, B_2 and B_5 were respective regression coefficients of these variables estimated through Model.* The result is presented in Table 7.3.

As mentioned earlier, gross cropped area, irrigated area and rainfall explained nearly 62.1 per cent variation in foodgrain production in North Bihar during the study period (1970-71 to 1993-94). During the period, the annual compound growth rate in foodgrain production was 2.70 per cent, whereas the growth rate estimated through the above model was 2.64 per cent.

Irrigated area emerged as the most important variable which contributed nearly 89.48 per cent to the growth in foodgrain production during the period under study. Land contribution towards growth in foodgrain production was 11.36 per cent indicating declining importance of land to the growth in foodgrain production in the project area. Rainfall, no doubt, exerted positive and significant influence on foodgrain production during the study period but its contribution to growth in foodgrain production was found negative but it was less than 1 per cent (-0.52%). It was probably due to a typical land situation and recurrent floods caused by higher rainfall. The findings particularly of rainfall indicated that the rainfall above the normal level would affect the growth of foodgrain production adversely.

* The model used is given in Chapter 3.

Table 7.3
Compound Growth Rate, Explained Growth Rate by Model and Contribution of Gross Cropped Area, Gross Irrigated Area and Rainfall to Growth Rate of Foodgrain [Output During PGRD (1970-71 to 1993-94)]*

Variables	Estimated Regression Co-efficients (b)	Growth Rate (per cent)	bx	Percentage Contribution
Gross Cropped Area (X_1)	0.99969*	0.3	0.299	11.36
	(0.58677)			
Gross Irrigated Area (X_2)	0.57473*	4.1	2.356	89.48
	(0.30147)			
Rainfall (X_5)	0.2183**	-0.1	-0.022	-0.84
	(0.06971)			
Explained growth rate of foodgrain output in (% per annum)		2.635		
Growth rate of foodgrain output (% per annum)		2.700		
				100.00

Figures in parentheses are the standard errors.

* Significant at 10 per cent level of probability.
** Significant at 5 per cent level of probability.

Among all these variables, irrigated area could further be increased by providing liberal assistance to farmers for creating irrigation structure in the project area which would help increasing the intensity of cropping and ultimately the gross cropped area. It is worth pointing out that the net area sown showed declining trend in North Bihar which would further decline in future due to shift of cultivable land to area put to non-agriculture use (Table 4.8). Hence, the increasing intensity is the only way to increase gross cropped area in North Bihar. Rainfall is still an important variable for foodgrain production in the project area but it is non-manipulative variable. At this juncture, it is prudent to study the behaviour of rainfall for providing feedback to biological scientists for formulating research programmes to evolve new varieties/seeds befitting the changed behaviour of rainfall.

Reference

Sharma, R.N. (1991), "Economics of Crop Sequences for Different Types of Land in Kalyanpur Block, Samastipur District", unpublished M.Sc. thesis, RAU, Pusa, p.94.

8

Summary and Conclusions

Indian agriculture moved from chronic food shortage in the early 1970s to food surplus status in the early 1990s. Almost all of the regions of the country have made impressive strides towards increasing per capita food production. But the spatial variations have been an important dimension of the spectacular growth in Indian agriculture caused by differences in agro-climatic situations, level of infrastructural facilities and inherent socio-economic characteristics of different regions of the country. Besides, the instability in farm output has also been an important aspect of agricultural growth over time. The fluctuations in agricultural production disturb the rhythm and mutual balances among many interdependent processes in the economy. Various studies were conducted to study the growth and instability in agriculture during pre- and post-green revolution periods but most of these relate to national/state level. There has been a dearth of such studies dealing with regional growth and instability in agriculture. Hence, the present study was undertaken to analyse the growth and instability in agricultural production along with the contributory factors affecting foodgrain production during post-green revolution period in North Bihar.

The study is primarily focussed to agricultural scenario of North Bihar, however, the zone-wise study has also been undertaken to have a comparative picture of regional agricultural growth/instability. The study covers the post-green revolution period, i.e., 1970-71 to 1993-94 which has categorised in three sub-periods, i.e., Period I (1970-71 to 1979-80), Period II (1980-81 to 1989-90),and Period III (1990-91 to 1993-94). Analysis is based on secondary data on area, production and productivity of principal crops along with different agro-economic variables affecting agricultural production in the project area. An effort has been made to present the data in different tables for drawing conclusions, however, various suitable statistical tools were also used to sharpen the relevant conclusions.

Analysis of data on areas under various crops indicated cereal dominated cropping pattern in North Bihar, since nearly three-fourth of gross cropped area was covered by cereals and showed increasing trend during period under study. The zone-wise analysis revealed the increasing proportion of area under cereals in the cropping pattern of the Zone I, but just reverse trend was observed in Zone II. It was probably due to shift in area from pulses to wheat cultivation in the Zone I.

Rice and wheat jointly constituted more than 60 per cent of gross cropped area in North Bihar during period under study. There was no marked increase in proportion of rice area but wheat showed increasing trend during the period under study. Despite the agricultural development, cereals dominated in the cropping pattern of North Bihar which was more pronounced during the early nineties. Oilseeds are, no doubt, unimportant crops with respect to area but gained coverage during the period under study. On the other hand, pulses are important non-cereal food crops but witnessed declining share in gross cropped area in both the zones during period under study.

During post-green revolution period, wheat emerged as an important cereal crop in North Bihar which occupied nearly one fourth of gross cropped area in the year 1994. The increasing share of wheat in gross cropped area was more pronounced in Zone I. Maize has been an important crop in the project area but its share in gross cropped area declined marginally during the period under study. It is worth pointing out that the area under kharif maize declined, however, area under rabi maize showed spectacular increase during the period under study in the project area (Appendix III). There was marked decline in area under barley, during post-green revolution period. *Arhar*, lentil, gram, rape and mustard and sugarcane did not show any specific trend, particularly over three points of time under study.

The coefficient of concordance test revealed that there was significant change in cropping pattern in the project area over three specified points of time, i.e., triennium ending 1974, 1984 and 1994. Zone I and Zone II also differed with respect to cropping pattern at three points of time. Hence, it may be said that there has been significant change in cropping pattern in the project area during the post-green revolution period. It has probably been contributed by bio-chemical and genetic innovations in principal crops like, wheat, rice and maize during the post-green revolution period. However, the regional difference in the cropping pattern was found, probably due to difference in ecosystem, infrastructural facilities and socio-economic milieu of the zones under study.

To study the changes in area, production and productivity of principal crops, the period of study was divided into two subperiods, i.e., early period of green revolution (1974 to 1984) and latter period of green revolution (1984 to 1994). It was observed that the area under cereals and oilseeds recorded an increase in both the periods. However, decadal increase in area was comparatively higher under oilseeds (9.29% and 11.51%) than that of cereals (5.34% and 6.18%). The increase in area under cereals was mainly due to increase in area under rice and wheat during early period of green revolution, whereas the corresponding increase in the latter period of green revolution was only due to increase in area under wheat (2.40 lakh ha). Area under maize declined continuously but decadal decline was comparatively higher during early period of green revolution (-39.26%) than the decline observed during latter period of green revolution (–5.56%). It is worth pointing out that the decline in maize area was only due to decline in area under kharif maize but the adoption of rabi maize at large scale during latter period of green revolution arrested the declining trend in area under maize.

Pulses witnessed a decadel increase of 4.59 per cent in area during early period of green revolution but a decadel decline of 3.90 per cent was observed in latter period of green revolution. The above phenomenon was attributed by the decline in area under major pulse crops like, gram and *arhar* during the period under study. Since area occupied by these crops has probably been shifted to wheat cultivation. Moreover, there was an increase in area under pulses in Zone II, during early period of green revolution also which was probably due to slow spread of wheat technology in the zone during early period of green revolution.

In North Bihar, production of all principal crops, viz., rice, wheat, pulses and oilseeds increased during the period under study. But rice, pulses and oilseeds recorded comparatively higher increase during early period of green revolution than that of latter period of green revolution. On the other hand, wheat and maize recorded comparatively higher rate of increase in their production during latter period than that of early period of green revolution. Hence, the general trend of declining production growth rate of foodgrains in the country does not hold true for North Bihar, particularly for wheat and maize. It may be noted that the production of wheat exceeded the production of traditional principal crop of rice in North Bihar during early period of nineties. Despite the decline in area under maize during both the periods, that is early and latter periods of green revolution under study, the production of maize increase by 22.30 per cent and 59.08 per cent respectively. It was made possible due to adoption of winter maize in the area.

The zone-wise analysis indicated a decline of 2.24 per cent in rice production in Zone II but moderate growth in Zone I during latter period of green revolution. It was only due to decline in area under rice in Zone II and a marginal increase in rice area in Zone I during latter period of green revolution. Moreover, there was comparatively slower growth in rice productivity in both the zones during the period. The above findings warranted location specific research in rice cultivation to enhance the rice productivity in the project area, in general, and in Zone II, in particular.

Analysis of compound annual growth rate in area, production and productivity of principal crop categories/crops during study period (1970-71 to 1993-94), Period I (1970-71 to 1979-80), Period II (1980-81 to 1989-90) and Period III (1990-91 to 1993-94) indicated an increase in area under all the principal crop categories, not only in North Bihar but both the zones under study. The area under foodgrains recorded comparatively higher growth rate (0.6%) than the corresponding growth recorded for cereals (0.4%), pulses (0.1%) and oilseed (0.4%) during study period in the project area. The sub-period-wise analysis indicated a decline in area growth rate of cereals and pulses from Period I to Period II which again increased during Period III. On the other hand, area growth rate of foodgrains increased from Period I (2.0%) to Period II (2.9%) but declined during Period III (0.6%). Area growth rate of oilseeds declined continuously over the specified sub-period under study. The Zone I witnessed almost the similar trends in the area growth rate of principal crop categories but area under foodgrains recorded much higher growth in the Zone I during Period II. On the other hand, Zone II recorded much higher growth rate in area under cereals, pulses and foodgrains during Period III which was only due to late adoption of modern technology in crop cultivation in the latter zone.

The production growth rates of all the principal crop categories were positive in North Bihar during study period and sub-periods under study. Between sub-periods, the production growth rates of all the principal crop categories were much higher during Period III than that of Period I and Period II. The zone-wise analysis revealed that Zone II performed much better in cereals, pulses and foodgrain production during Period III than that of Zone I. Growth rate in pulses and oilseeds production also increased from Period II to Period III in Zone I. The growth in production of major cereals (rice and wheat) was due to increase in both area and productivity, whereas increase in pulses production was attributed mainly to increase in productivity during study period in general and during latter period of green revolution in particular (Appendix VII). The zone-wise analysis revealed that the productivity played a major role in increasing production of crop categories in Zone II, specially during Period III, whereas, there was a deceleration in productivity growth of foodgrain crops in Zone I which affected the production of these crop categories. Moreover, Zone I performed much better in increasing production of oilseeds through enhancing productivity.

In North Bihar, rice production, no doubt, increased during the study period which was mainly due to increase in productivity. During Period I and Period II, rice recorded much higher growth in production but experienced a decline in production during Period III. Despite the growth in area under rice during Period III, rice production showed a declining trend due to a decline in rice productivity in the project area. Both the zones under study experienced deceleration in the growth rate of rice production over the sub-periods under study, however, the performance of Zone II was comparatively better than Zone I. On the basis of above discussion, it may be said that there is no scope to increase the area under rice in North Bihar, hence, the rice production can be increased only through increasing productivity.

Wheat is the only crop which recorded a positive growth rate in area, production and productivity across the zone over different sub-periods under study. There was a deceleration in area, production and productivity during Period II which picked up during Period III. It was probably due to unfavourable economics of wheat cultivation during eighties which was caused by comparatively lower price of wheat in comparison to competing crops during the period. It is worth pointing out that the sharp increase in wheat prices since 1991 might have motivated farmers to shift area from pulses to wheat crop, particularly, during the latter period of green revolution.

Despite the decline in area under maize during the period under study, production showed a significant increase due to significant increase in productivity. The crop performed much better during Period III and its area, production and productivity recorded significant increase in North Bihar. Zone II, no doubt, witnessed a decline in area during Period II which increased by nearly 9 per cent during Period III. As discussed earlier, dynamics of maize production is influenced by extent of adoption of rabi maize in the project area.

On the whole, North Bihar experienced a decline in area under major pulses during post-green revolution period which was mainly contributed by comparatively a higher decline in area in Zone I. The Zone II performed comparatively better through bringing area under pulses, particularly during early period of nineties. Arhar and gram registered declining growth in their production but lentil recorded a growth of 2.2 per cent during study period. The decline in production growth rate of pulses was mainly due to decline in area of these crops during post-green revolution period, which was mainly contributed by higher decline in area in Zone I. Principal pulse crops did not show any trend in productivity however, lentil recorded significant increase in productivity during period under study. Hence, it may be said that the pulses production can only be increased by increasing the coverage since productivity did not show any recovery, probably due to their cultivation on marginal land, lack of appropriate pulses technology and poor implementation of programme of pulses production in the project area.

Rape and mustard production also showed an increasing trend during the period under study, but got a setback during Period II which picked up during Period III. The Zone I performed comparatively better in oilseed production during the period under study. The increasing trend in oilseed production was again contributed by the productivity growth rate caused by better adoption of improved package of practices and seeds by farmers.

Sugarcane production also increased during the study period but got a setback during Period I due to decline in area and productivity which was probably caused by unfavourable economics of sugarcane cultivation in comparison to competing paddy-wheat crop sequence. The sugarcane production picked up during Period II and Period III, probably due to increase in productivity on the one hand and an increase in administered price of sugarcane, on the other.

On the basis of above discussion, it may be concluded that

there has been a steady growth in production of almost all the crops though at different rates but it has not been smooth for food grain crops in general and principal kharif crops in particular.

The study of instability indicated that the principal foodgrains like, wheat, maize and arhar witnessed a continuous decline in instability over three sub-periods under study. But rice witnessed sharp increase in stability from Period I to Period II which declined during Period III. It is worth pointing out that wheat had the highest instability during the Period I (50.04%) which declined to almost at the lowest level (6.94%) during period III. During the early period of green revolution, wheat production increased due to an increase in both area and productivity which was accompanied with higher level of instability, but wheat production growth rate experienced deceleration in production growth rate and instability during latter period of green revolution.

The instability in oilseed production also witnessed an increase from Period I to Period II but became more stable during Period III. Barley observed a decline in production due to a decline in area and witnessed a decline in instability also. It is also worth pointing out that the instability in area and productivity of almost all crops move in the same direction and their increasing/decreasing trend resulted in a decline in instability. Hence, it may be said that the increase in production of a particular crop due to a spectacular increase in area and productivity would accompany the increase in instability also, but an increase in production largely due to the increase in productivity would help declining production instability of particular crop.

An effort was also made to study the relationship between production growth and instability of principal crops and it was found that the rice did not have significant relationship between production growth and instability. Wheat, no doubt, had positive and significant relationship between growth and instability in production during latter period of green revolution. However, it has a negative relationship during early period of green revolution when wheat production experienced a spectacular increase. The positive relationship between growth and instability in wheat production was found during latter period of green revolution when wheat production growth rate experienced a decleration along with a decline in instability in production. Hence, the instability in rising production trends during early period of adoption of new technologies may be considered as teething trouble in adoption of new technology. Production instability starts declining just after getting the particular technology stabilised. The relationships between production growth and instability were also found positive and significant for maize during the study period and early period of green revolution. Whereas lentil had positive relationship during the study period, early and latter periods of green revolution. But rape and mustard had negative and significant relationship between growth and instability during latter period of green revolution. The above findings clearly indicated that the increase in production mainly due to an increase in productivity had positive relationship but the increase in production due to an increase in area and productivity had negative relationship between growth and instability in production of a particular crop. Hence, it may be said that the stability in production may be achieved by increasing productivity without a decline in area in a particular crop. It could only be achieved by making investment in agricultural research and keeping favourable agricultural price policy.

The components of foodgrain production have been analysed and it was found that land is still exerting a significant influence on foodgrain production in North Bihar. Irrigation and rainfall also emerged as important components of foodgrain production in the project area. All the three significant variables (gross cropped area, gross irrigated area and annual rainfall) could explain only 62.1 per cent variation in foodgrain production, however, rainfall is still in position to explain nearly 24.90 per cent of variation in foodgrain production in the project area. Gross cropped area is also able to explain 20.91 per cent variation in foodgrain production, indicating importance of land in foodgrain production. Irrigation, no doubt, is an important factor of production but it has explained only 16.29 per cent variation in foodgrain production. It was probably due to poor irrigation infrastructure in the project area. The inefficient canal system, non-functioning of public tubewells and inadequate availability of fuel (diesel) might be responsible for comparatively lower contribution of irrigation to foodgrain production in the project area.

The components of growth in the foodgrain production were also estimated to have an idea about the influence of individual factor in increasing foodgrain production in the project area. Irrigated area emerged as the most important variable which contributed nearly 89.48 per cent to the growth in foodgrain production during the period under study. Land contribution towards growth in foodgrain production was 11.36 per cent indicating declining importance of land to the growth in foodgrain production in the project area. Rainfall, no doubt, exerted positive and significant influence to foodgrain production during the study period but its contribution to growth in foodgrain production was found negative (-0.52%). It was probably due to a typical land situation and recurrent floods caused by higher rainfall. The findings, particularly of rainfall, indicated that the rainfall above the normal level would affect the growth of foodgrain production adversely in North Bihar.

Among all these variables, irrigated area could further be increased by providing liberal assistance to farmers for creating irrigation structure in the project area which would help increasing the intensity of cropping and ultimately the gross cropped area. It is worth pointing out that the net area sown showed a declining trend in North Bihar which would further decline in future due to shift of cultivable land to area put to non-agriculture use (Table 4.8). Hence, the increasing intensity is the only way to increase gross cropped area in the project area. Rainfall is still an important variable for foodgrain production in the project area but it is a non-manipulative variable. At this juncture, it is prudent to study the behaviour of rainfall for providing feedback to biological scientists for formulating research programmes to evolve new varieties/technologies befitting the changed behaviour of rainfall.

On the basis of above discussion, it may be concluded that the cropping pattern of the project area is still predominated by cereal crops. Among the cereals, wheat is emerging as major cereal crop in the project area. There has been a significant change in the cropping pattern during post-green revolution period. The zone-wise analysis also indicated a significant regional difference in cropping pattern since early period of green revolution. The predominance of cereal crops in the cropping pattern may be due to foodgrain scarcity in the project area. However, the regional differences in cropping pattern has been caused by regional variations in agro-climatic situations.

All the principal crops recorded growth in production but the performance of rice---the principal staple food crop in the project area---has not been impressive due to almost stagnant growth in area and insignificant increase in productivity. It is worth-mentioning that 60 per cent of the rice area is under unfavourable ecosystem in the project area for which appropriate technologies are still to be evolved. Wheat performed much better in the project area and its production exceeded the rice production in early nineties, but there has been a deceleration in rice production and productivity growth rates during the latter period of green revolution. Any improvement in wheat production can again be achieved by evolving better varieties and by improving irrigation system in the project area. Maize also performed better in the project area, mainly due to adoption of winter maize on large scale. Pulse production witnessed a setback due to a decline in area but oilseeds performed much better during the period under study mainly due to an increase in productivity. Moreover, special efforts are needed to evolve the varieties of pulses and oilseeds for increasing their production to meet the requirements in the project area. The sugarcane production also got a setback in the early period of green revolution, but picked up during latter period due to an increase in productivity and favourable price policy.

Almost all the crops witnessed an increase in instability during early period of green revolution which got almost stabilised during latter period of green revolution. There has been no clear-cut evidence of relationship between growth and instability but declining growth rates in production and instability showed positive and significant relationship for wheat, maize and lentil which does not support the hypothesis that the increase in production growth rate would be accompanied by increase in instability. Oilseeds had negative and significant relationship between growth rate and instability in production which was caused by increasing production due to an increase in area and productivity. Hence, it may be said that the increase in area under crop along with productivity would stabilise the production of a particular crop.

Land, irrigation and rainfall were found exerting significant influence on foodgrain production. Among these three variables, irrigated area can only be increased since there is no scope of increasing the land area and rainfall. However, an increase in irrigated area would help increasing the intensity of cropping and ultimately the gross cropped area. The increase in irrigational infrastructure would also be able to conserve the rain water for rational utilisation at appropriate time.

While examining the components of growth in foodgrain production, irrigation again emerged as an important factor which contributed nearly 89.48 per cent to the growth of foodgrain production. Hence, it may be said that an increase in investment on irrigation is the necessary factor for increasing agricultural production in the project area. Moreover, investment on agricultural research is also required for evolving varieties and package of practices for increasing foodgrain production because the growth in foodgrain production would only be feasible through increasing productivity of principal crops.

Appendices

Appendix I

Percentage Decline in Output of Foodgrains in Drought Years/Increase in Bumper Years as Compared to Preceding Years

Drought Years	*Percentage Change*
1979-80	-17.16
1982-83	-5.42
1987-88	-13.19
1990-91	-8.91
Bumper years	Percentage change
1975-76	14.32
1984-85	9.36
1988-89	39.26

Appendix II

Zone-wise Area of Different Crops in North Bihar During 1995-96

(area in ha)

Crop/Crop Categories	*Zone I*	*Zone II*	*North Bihar*
Rice	1361500	814000	2175500
	(45.50)	(41.97)	(44.11)
Wheat	821986	394578	1215564
	(27.47)	(20.29)	(24.64)
Maize	237757	214276	452033
	(7.94)	(11.04)	(9.16)
Barley	9275	2133	11408
	(0.30)	(0.10)	(0.23)
Other Cereals	87572	114618	202190
	(2.92)	(5.90)	(4.09)
Total Cereals	2518090	1538605	4056695
	(84.15)	(79.34)	(82.26)
Arhar	24462	4331	28793
	(0.81)	(0.22)	(0.58)
Gram	10546	6633	17179
	(0.35)	(0.34)	(0.34)
Lentil	32691	18533	51224
	(1.09)	(0.95)	(1.03)
Other Pulses	115134	144470	259604
	(3.84)	(7.44)	(5.26)
Total Pulses	182833	173967	356800
	(6.11)	(8.97)	(7.23)
Rape and Mustard	35060	41880	76940
	(1.17)	(2.15)	(11.56)

Contd...

Contd...

Other Oilseeds	28637 (0.95)	20423 (1.05)	49060 (0.99)
Total Oilseeds	63697 (2.12)	62303 (3.21)	126000 (2.55)
Sugarcane	105793 (3.53)	2887 (0.14)	108680 (2.20)
Miscellaneous Crops	121687 (4.06)	161644 (8.33)	283331 (5.74)
Gross Cropped Area	2992100	1939406	493506

Figures in parentheses indicate the percentage to gross cropped area.

Appendix III

Area and Production of Bhadai Maize, Rabi Maize and Summer Maize at Three Points of Time (1980-81, 1984-85 and 1991-92)

Zone	Points of Time	Bhadai Maize		Rabi Maize		Summer Maize	
		Area (in ha)	Production (in tonnes)	Area (in ha)	Production (in tonnes)	Area (in ha)	Production (in tonnes)
I							
	1980-81	254084	194891	76359	151886	38851	57890
	1984-85	182162	208155	86647	229094	39874	72287
	1991-92	143188	265112	130670	355719	38563	79469
II							
	1980-81	54967	44371	12554	29589	65160	46736
	1984-85	37852	51750	16479	25673	55009	66536
	1991-92	41469	57731	51081	139758	66612	131650

Appendix IV

Area, Production and Yield of Rice in North Bihar During Study Period (1970-71 to 1993-94)

Year	*Area (in lakh ha)*	*Production (in lakh tonnes)*	*Yield kg/ha*
1970-71	17.54	13.52	770
1971-72	21.37	11.56	540
1972-73	18.99	14.21	750
1973-74	19.10	13.16	690
1974-75	20.93	15.61	750
1975-76	20.62	15.81	770
1976-77	22.35	22.08	990
1977-78	20.80	17.67	850
1978-79	21.94	18.82	860
1979-80	20.39	18.09	850
1980-81	21.72	19.81	910
1981-82	21.32	16.27	760
1982-83	18.85	12.64	670
1983-84	28.52	20.98	700
1984-85	21.07	19.04	900
1985-86	23.22	22.42	970
1986-87	22.24	22.09	990
1987-88	20.12	16.00	800
1988-89	22.02	25.67	1170
1989-90	22.22	23.81	1070
1990-91	20.17	20.11	1000
1991-92	22.11	22.76	1030
1992-93	18.34	11.32	620
1993-94	21.36	21.75	1020

Appendix V

Area, Production and Yield of Wheat in North Bihar During Study Period (1970-71 to 1993-94)

Year	*Area (in lakh ha)*	*Production (in lakh tonnes)*	*Yield kg/ha*
1970-71	5.01	6.24	1250
1971-72	6.43	7.98	1240
1972-73	8.55	12.08	1410
1973-74	8.19	7.96	970
1974-75	8.62	12.38	1440
1975-76	9.70	12.28	1270
1976-77	10.32	19.47	1890
1977-78	9.55	28.20	2950
1978-79	9.32	11.93	1280
1979-80	8.98	10.75	1200
1980-81	9.06	12.32	1360
1981-82	8.53	12.07	1410
1982-83	8.85	12.86	1450
1983-84	9.54	14.54	1520
1984-85	9.69	16.39	1690
1985-86	10.16	15.85	1560
1986-87	9.67	14.39	1490
1987-88	11.23	16.74	1490
1988-89	11.66	20.09	1720
1989-90	10.84	17.95	1660
1990-91	10.36	19.52	1880
1991-92	11.56	20.93	1810
1992-93	11.75	22.47	1910
1993-94	11.50	22.72	1910

Appendix VI

Area, Production and Yield of Maize North Bihar During Study Period (1970-71 to 1993-94)

Year	*Area (in lakh ha)*	*Production (in lakh tonnes)*	*Yield kg/ha*
1970-71	1.07	5.26	4880
1971-72	10.87	2.82	260
1972-73	4.20	4.82	1150
1973-74	15.32	5.66	370
1974-75	4.93	7.01	1420
1975-76	4.63	4.89	1060
1976-77	4.10	6.22	1520
1977-78	5.31	6.96	1320
1978-79	5.20	6.23	1200
1979-80	4.90	5.28	1080
1980-81	5.04	5.25	1040
1981-82	4.51	4.88	990
1982-83	4.69	6.73	1430
1983-84	4.47	5.85	1310
1984-85	4.08	6.47	1590
1985-86	3.86	5.76	1490
1986-87	3.86	6.63	1720
1987-88	3.39	8.74	2570
1988-89	4.31	7.59	1760
1989-90	4.53	8.98	1980
1990-91	4.22	7.73	1830
1991-92	4.64	10.29	2220
1992-93	4.57	8.80	1920
1993-94	4.61	9.31	2020

Appendix VII

Variation in Production of Principal Crops Due to Change in Area and Productivity in North Bihar During Early and Latter Period of Green Revolution

(in per cent)

Crop	*Due to Change Area*		*Due to Change in Productivity*	
	1974-84	*1984-94*	*1974-84*	*1984-94*
Rice	27.91	-9.55	72.09	109.55
Wheat	64.38	53.05	35.61	46.95
Maize	-174.81	-9.58	274.81	109.58
Total Cereals	32.07	19.48	67.92	80.51
Total Pulses	17.91	-72.50	82.09	172.50
Total Oilseeds	25.35	39.47	74.65	60.51
Sugarcane	-51.91	49.58	-48.07	50.42

Bibliography

Abraham, T.P. and Raheja, S.K. (1967), "An Analysis of Growth of Production of Rice and Wheat Crops in India", *Indian Journal of Agricultural Economics* 22(3): 1.

Bandyopadhyay, A. (1989), "Growth and Instability in the Production of Main Cereal Crops of West Bengal and Punjab-Haryana, 1950-51 to 1984-85: A Note", *Indian Journal of Agricultural Economics* 44 (2): 168.

Bhagat, L.N. (1982), "Growth Rates of Output of Coarse Grains and Pulses in a Backward Economy: A Study of Temporal Spatial Variation in Chotanagpur", *Indian Journal of Agricultural Economics* 37 (3): 396.

Bhat, G.M., Dhar, M.K. and Khan, G.M. (1986), "Study of Growth Rates of Major Crops in Jammu and Kashmir", *Agricultural Situation in India* 41 (7): 559.

Bhát, G.M., Dhar, M.K., Baig, B.A. and Zutshi, S.N. (1989), "Crop Concentration and Cropping Pattern in Jammu & Kashmir States", *Agricultural Situation in India* 43 (2): 57.

Chandra, D.R., Tiwari, C.B. and Singh, G.N. (1983), "Vertical Growth for Agriculture: A Regional Analysis of Uttar Pradesh", *Agricultural Situation in India* 38(6): 389.

Chattopadhyaya, Alok (1991), "Growth and instability in Agriculture: The Case of Rice in West Bengal", *Recent Advances in Agricultural Statistics Research*, New Delhi, Wiley Eastern, p.118.

Chawla, J.S., Gill, S.S. and Dhariwal, M.S. (1973), "Evaluation of Minor Irrigation Works in District Amritsar", *Indian Journal of Agricultural Economics* 30 (3): 244.

Cummings, R.W. and Ray, S.K. (1969), "1968-69 Foodgrain Production: Relative Contribution of Weather and New Technology", *Economic and Political Weekly*.4(39): A-163.

Dasgupta, K.R. (1965), "Cropping Pattern and Utilization of Land in Assam 1955-56 to 1964-65", *Agricultural Situation in India* 20 (4): 215.

Desai, B.M. (1977), "Cropping Pattern of Farm Families", *Indian Journal of Agricultural Economics* 27 (1): 51.

Deshpande, R.S. (1991), "The Growth Instability Nexus: A Study of Maharashtra", *The Recent Advances in Agricultural Statistics Research*, New Delhi, Wiley Eastern, p. 100.

Dhawan, B.D. (1991), "How Stable is Indian Irrigated Agriculture?", *Recent Advances in Agricultural Statistics Research*, New Delhi, Wiley Eastern, p. 53.

Gangawar, A.C. and Pandey, R.N. (1982), "Stagnation in Production of Pulses: An Economic Analysis", *Agricultural Situation in India* 37(5): 283.

George, P.S. and Mukherjee, Chandan (1986), "A Disaggregated Analysis of the Growth Performance of Rice in Kerala", *Indian Journal of Agricultural Economics* 41(1):1.

Giri, R., Sastri, A.V.K. and Somajulu (1966), "Components of Crop Output Growth in India", *Indian Journal of Agricultural Economics* 21 (4): 185.

Grewal, S.S. and Rangi, P.S. (1983). "An Analytical Study of Growth of Punjab agriculture", *Indian Journal of Agricultural Economics* 38(4): 517.

Guha, G.S. and Mathur, N. (1995), "Output, Employment and Poverty: A Situation Analysis Across Agro-climatic Regions", Seminar on Agricultural Growth, Employment and Rural Poverty in Rural India held at A.N. Sinha Institute, Patna, December 8-10.

Hanumantha Rao, C.H. and Gulati, A. (1994), "Indian Agriculture Emerging Perspectives and Policy Issues", *Economic and Political Weekly*, 29(53): A-158.

Hanumantha Rao, C.H., Ray, Susanta K. and Subbarao, K. (1988), *Unstable Agriculture and Droughts*, New Delhi. Vikas Publishing House.

Haffis, S., Rama Rao, C.A., Reddy, Y.V.R. and Katyal, J.C. (1994), "Growth, Instability and Drought Sensitivity of Sorghum and Pearl Millet Production", *Agricultural Situation in India* 47 (10): 97.

Hazell, Peter B.R. (1982), "Instability in Indian Foodgrain Production", International Food Policy Research Institute, *Research Report* No.30, Washington D.C.

Jha, B. K. (1994). "Growth and Instability in Agriculture Associated with New Agricultural Technology: District Level Evidence", *Agricultural Situation in India* 49(7): 517.

Krishna, R. and Tyagi, B.N. (1973), "Study of Growth Rate of Area, Production and Productivity of Wheat Crop in Uttar Pradesh", *Agricultural Situation in India* 28 (2): 83.

Kumar, A. and Kaul, S. (1991), Estimation of Contribution of Growth Components and Instability in agriculture", *Recent Advances in Agricultural Statistics Research*, New Delhi, Wiley Eastern, p. 75.

Lal, R. and Singh, R.P. (1994), "Structural Change in Haryana Agriculture: The Issue of Growth and Equity", *Agricultural Economics Research Review* 7 (2): 129.

Mahendra Dev, S. (1987), "Growth and Instability in Foodgrain Production", *Economic and political weekly.* 22 (39): A-82.

Malik, H.S., Newas, S. and Gangwar, A.C.(1988), "Production and Marketing of Wheat in Haryana", *Research Bulletin* No. 21, Department of Agricultural Economics, H.A.U., Hissar.

Mehra, Shakuntala (1981), "Instability in Indian Agriculture in the Context of the New Technology", International Food Policy Research Institute, *Research Report*, No.25, Washington, D.C.

Mohammad, Ali Khan, N. and Singh, T. (1979), *Regional Variations in Agricultural Productivity: A Study of Bihar in Dynamics of Agricultural Development*, (ed.) Ali Mohammad, Delhi.

Mukhopadhyaya, S.K. (1976), *Sources of Variation in Agricultural Productivity: A Cross-Section Time Series Study in India*, Delhi.

Myers, R.H. (1971), *Response Surface Methodology*, Boston Allyn and Bacon, p. 48.

Pal, S. and Sirohi, A.S. (1991), "Adoption of HYV Seeds and Growth and Instability in Crop Production in India: Implications for Policy", *Recent Advances in Agricultural Statistics Research*, New Delhi, Wiley Eastern, p. 81.

Pal, Suresh and Sirohi, A.S. (1988), "Sources of Growth and Instability in the Production of Commercial Crops in India", *Indian Journal of Agricultural Economics*, 43(3): 456.

Parikh, A. (1966), "State-wise Growth Rate in Indian Agricultural Output: An Econometric Analysis", *Artha Vijanana* 8 (1): 1-52.

Paroda, R.S. (1995), "Sustainable Growth: Major Plank of Strategy", *Survey of Indian Agriculture*, p. 17.

Prasad, A.K. (1987), "A study on Trend in Area, Production and Productivity of Pulses in Bihar", Unpublished thesis, Department of Agricultural Economic, Rajendra Agricultural University, Pusa, Samastipur, p.68.

Raghavarao, D. (1983), *Statistical Techniques in Agricultural and Biological Research*, New Delhi, Oxford and IBH, p. 338.

Raj, K.N. (1979), "Economic Trends and Changes in Strategy", *Mainstream* 18(16): 15.

Ram, R. (1979), "Technological Determinants and Agricultural Development: An Analysis of Regional Variations in Eastern Uttar Pradesh", Ali Mohammad (ed.), *Dynamics of Agricultural Development in India*, Delhi.

Ramasubban, T.A. (1963), "Some Statistical Measures to Determine Changes in Cropping Pattern", *Agricultural Situation in India* 17(11): 1153.

Randhawa, N.S. and Dev, G. (1972), "Fertilizer Usage in Changing

Agriculture in Punjab", *Indian Farming* 22(6): 76.

Rao, M.V. (1986), "Agricultural Production: Pace and Pattern of Growth", *Indian Agricultural Development Since Independence*, New Delhi, Oxford & IBH, p.95.

Rath, Nilakanth (1980), "A Note on Agricultural Production in India During 1955-78", *Indian Journal of Agricultural Economics.* 35 (2): 91.

Ray S.K. *et al.* (1985), "Impact of Changes in Cropping Pattern in Farm Income", *Agricultural Situation in India.* 40(7): 625.

Ray, S.K. (1983), "An Empirical Investigation of the Nature and Causes for Growth and Instability in Indian Agriculture", *Indian Journal of Agricultural Economics.* 38 (4): 459.

Ray, S.K.(1991), "Instability in Indian Agriculture Revisited", *Recent Advances in Agricultural Statistics Research*, New Delhi Wiley Eastern, p. 11.

Sagar, Vidya (1978), "Contribution of Individual Technological Factors in Agricultural Growth: A Case Study of Rajasthan", *Economic and Political Weekly* 13(25): A-63.

Sen, S.R. (1967), "Growth and Instability in Indian Agriculture", *Agricultural Situation in India* 21 (10): 827.

Sharma, R.N. (1991), "Economics of Crop Sequences for Different Types of Land in Kalyanpur Block, Samastipur District", Unpublished M.Sc. Thesis, Deptt. of Agril. Economics, R.A.U., Pusa, Samastipur. p. 94.

Singh, Chhotan and Mathur, V.C. (1994), "Growth and Instability in the Production and Price of Potato in India", *Agricultural Situation in India.* 49(6): 429.

Singh, Daroga (1981), "Imbalances in Agricultural Growth", *Indian Journal of Agricultural Economics.* 36(1): 75.

Singh, Darshan (1966), "Review of Cropping Pattern and Land Utilization in Punjab During Plan Periods", *Agricultural Situation in India.*21 (2 & 3):85.

Singh, H., Tomar, B.S. and Suhag, L.R. (1982), "Growth Rates and Trends in Area, Production and Productivity of Oilseed Crops in Haryana", *Indian Journal of Agricultural Economics.* 37(3): 399.

Singh, I.J. (1989), "Agricultural Instability and Farm Poverty in India", *Indian Journal of Agricultural Economics* 44(1): 1.

Singh, J.P. and Gangwar, A.C. (1991), "Instability in Cereal Production in Haryana: A Decomposition Analysis", *Recent Advances in Agricultural Statistics Research*, New Delhi Wiley Eastern, p. 130.

Singh, M. (1980), "Agricultural Development Differentials in States of Haryana and the Punjab: A Comparative Study", Unpublished Ph.D. thesis, Division of Extension Education, IARI, New Delhi, p. 125.

Singh, S.P. and Gangwar, A.C. (1986). "Trends and Variability in Area, Production and Productivity of Coarse Grains and Pulses in Haryana", *Agricultural Situation in India* 41(2):81.

Sinha, D.K. (1991), "An Evaluation of Technological Progress in Bihar Agriculture with Special Reference to Rice: An Econometric Analysis)", Unpublished Ph.D. thesis, Deptt. of Agril. Economics, Rajendra Agricultural University, Pusa, Samastipur.

Sirohi, A.S., Vasisht, A.K. and Singh, Chhotan (1983), "Trend of Agricultural Productivity and Production in India", *Agricultural Situation in India.* 38(8): 293.

Sridharan, B. and Radhakrishnan, A.S. (1978), "A Study on the Factors Affecting Changes in the Cropping Pattern in Nilgir District of Tamil Nadu", *Indian Journal of Agricultural Economics* 33(4): 14.

Swant, S.D. and Achuthan, C.V. (1995), "Agricultural Growth Across Crops and Regions: Emerging Trends and Patterns", *Economic and Political Weekly.* 25 March, 1995, p. A-2.

Vaidyanathan, A. and Mukherjee, C. (1991), "State-wise Analysis of Agricultural Growth", *Recent Advances in Agricultural Statistics Research,* New Delhi, Wiley Eastern, p. 34.

Venkatramanan, L.S. and Prahaladhachar, M. (1981), "Growth Rates and Cropping Pattern Changes in Six States: 1950 to 1975", *Indian Journal of Agricultural Economics.* 35(2): 71.

Venkata Rao, N. (1991), "Trends in Growth and Instability in Agriculture in North Eastern India", *Recent Advances in Agricultural Statistics Research*, New Delhi, Wiley Eastern, p. 110.

Index